MONOGRAPHS ON PHYSICAL BIOCHEMISTRY

GENERAL EDITORS

W. HARRINGTON   A. R. PEACOCKE

THE OSMOTIC PRESSURE OF BIOLOGICAL MACROMOLECULES

BY

M. P. TOMBS

SENIOR SCIENTIST, UNILEVER RESEARCH LABORATORIES, COLWORTH HOUSE

AND

A. R. PEACOCKE

DEAN OF CLARE COLLEGE, CAMBRIDGE

CLARENDON PRESS · OXFORD

1974

Oxford University Press, Ely House, London W.1

GLASGOW NEW YORK TORONTO MELBOURNE WELLINGTON
CAPE TOWN IBADAN NAIROBI DAR ES SALAAM LUSAKA ADDIS ABABA
DELHI BOMBAY CALCUTTA MADRAS KARACHI LAHORE DACCA
KUALA LUMPUR SINGAPORE HONG KONG TOKYO

ISBN 0 19 854606 8

© OXFORD UNIVERSITY PRESS 1974

All rights reserved. No part of this publication may be reproduced, stored in a retrieval system, or transmitted, in any form or by any means, electronic, mechanical, photocopying, recording or otherwise, without the prior permission of Oxford University Press

PRINTED IN NORTHERN IRELAND AT THE UNIVERSITIES PRESS, BELFAST

PREFACE

AN IMPORTANT strand in the history of the study of the physical chemistry of biological macromolecules was the gradual recognition that biological 'colloids' were in fact macromolecules. This fundamental concept was established largely by studying solutions of such macromolecules under two particular kinds of equilibrium conditions. In one of these—'osmotic' equilibrium with which this monograph is concerned—an aqueous solution of the macromolecule contained inside a membrane, permeable only to water and salts, the 'solvent', and not to the macromolecule, is brought into equilibrium with respect to the diffusible species. This is achieved by application to the solution of an extra pressure, called the 'osmotic pressure'. The other phenomenon, whose study by Svedberg during the 1920s and 1930s was historically significant in establishing the character of biological macromolecules, is the equilibrium re-distribution of concentration which occurs when a column of its solution is subjected to a very large gravitational field, as in the ultracentrifuge.

Since in both of these situations the macromolecular systems are at equilibrium, the powerful and comprehensive ideas developed to describe the thermodynamics of solutions are applicable to their description and to the formulation of the relationships between their observable parameters. Osmotic pressure and sedimentation equilibrium studies not only have been influential historically in guiding our understanding of the nature and interaction of biological macromolecules in solution, but they continue to be very important methods in their own right and there are indications that they may be of crucial importance for the examination of those proteins, especially enzymes, which dissociate into sub-units. Such dissociation is essentially a process of chemical equilibrium, albeit between much larger entities than is sometimes the concern of the chemist, and is therefore describable in terms of an equilibrium constant and its associated thermodynamic parameters. Hence methods which examine such systems at equilibrium, such as measurement of osmotic pressure and sedimentation equilibrium, are likely to prove particularly helpful and the importance of these methods therefore has wider import than is usually implied when they are regarded simply as one of the methods for determining molecular weights of macromolecules. Parameters dependent on molecular interaction are also, as will be seen later, obtainable from the study of these phenomena: these parameters have primarily a thermodynamic significance but are frequently capable of molecular interpretations which throw an interesting light on the nature of the interactions between the macromolecules themselves and between the macromolecules and other kinds of molecular species in the solution.

Osmotic pressure studies long preceded the advent of the ultracentrifuge and still occupied a key role in the armoury of the physical biochemist up to the 1940s, not only in actually characterizing biological macromolecules (notably proteins) by determining molecular weights, but also in developing precise ideas concerning the properties of their solutions and of their physiological function. For a period, such measurements became less fashionable, but the effectiveness of the method and the fundamental theoretical importance of the phenomenon was securely established during this period by the classical investigations of Scatchard and co-workers (1946, 1954). It is true that there was at that time some serious practical difficulties in making osmotic pressure measurements which outweighed their apparent simplicity and made other methods more attractive, notwithstanding their greater complications and expense. For example, osmotic pressure measurements needed relatively large amounts of material, and took at least several days, so that bacterial and fungal contamination was always a hazard. Membranes, too, were not easy to obtain and frequently required rather ill-characterized methods of preparation needing long application for success. All this has changed in the last few years, for little material is needed with modern equipment and measurements can be quick and sufficiently precise. Thus osmotic pressure measurement is now as accessible as any other method of macromolecular investigation and many, if not all, of its previous disadvantages have been overcome. In some situations it is the most powerful and informative method and this is particularly so when determination of number-average molecular weights is desirable, as in investigations of the sub-unit structure of proteins.

For the reasons already indicated the most pertinent theoretical framework for the descriptive and quantitative formulation of osmotic phenomena is afforded by the thermodynamic account of solutions in equilibrium. The first chapter is, therefore, devoted to an outline of this account, the formulation of suitable notations and development of the necessary thermodynamic relationships applicable to solutions of biological macromolecules at equilibrium both osmotic and otherwise. On this basis, the theory, both approximate and rigorous, of the osmotic pressure of solutions of biological macromolecules is elaborated in the second chapter. The important developments in the exposition of the thermodynamics of multi-component systems are set out and the connection between the thermodynamic and more molecular, statistical accounts of osmotic phenomena are taken into account.

Development of instrumentation has played a vital role in restoring osmotic pressure measurements as a useful tool for physical biochemists and polymer chemists and the third chapter gives a selective account of the different types of instrument available, together with sample calculations and information of a practical kind.

The final chapter reviews some previous applications of osmotic pressure measurements. It is striking that the problems have not changed much over the years though of course the gradual accretion of results from this and other methods has yielded a clearer picture of the protein molecule in solution. Investigations have been chosen for exposition because they illustrate a particular point or a particular method of interpretation and although many of the papers represent landmarks in the development of the subject there are other equally valuable papers we have not been able to include.

Our aim throughout has been to present the various facets of the osmotic pressure method so that its special virtues can be fully weighed when studying a particular biological macromolecule and choice has to be made from among the techniques available for elucidating the nature and interactions of such macromolecules in solution.

It is a pleasure to acknowledge the help which we have received from discussions with Dr. A. G. Ogston, F.R.S., President of Trinity College, Oxford, and from the advice of the staff of the Clarendon Press.

We also thank the following for permission to reproduce figures: the American Chemical Society (for Figs. 2.4, 3.2, 3.3, 3.8, 4.7, 4.9, 4.10); the Biochemical Journal (for Figs. 3.5, 3.11, 3.12, 4.6, 4.8); Acta Chemica Scandinavica (for Fig. 3.10); Comptes Rendues, Carlsberg Laboratories (for Figs. 4.2, 4.3, and 4.4); Journal of Biological Chemistry (for Figs. 4.11 and 4.12); and Makromolekular Chemie (for Figs. 3.7 and 4.5).

M.P.T.
A.R.P.

CONTENTS

1. THERMODYNAMICS OF MACROMOLECULAR SOLUTIONS — 1
2. THEORY OF OSMOTIC PRESSURE — 28
3. OSMOMETERS AND OSMOTIC PRESSURE MEASUREMENTS — 66
4. APPLICATIONS OF OSMOTIC PRESSURE MEASUREMENTS — 104

REFERENCES — 136

INDEX — 141

1
THERMODYNAMICS OF MACROMOLECULAR SOLUTIONS

WHEN a membrane which separates two fluid phases is impermeable to one (or more) of the components of a fluid mixture, there is a net flow of diffusible components across the membrane which is usually in such a direction that the non-diffusible component is diluted. This phenomenon, which is called *osmosis* (from the Greek for *push* or *impulse*), may be observed with gas mixtures of, say nitrogen and hydrogen, contained in a palladium vessel permeable only to hydrogen and immersed in nitrogen. Hydrogen flows across to the side free of that gas until its chemical potential is the same on both sides. But more usually the term refers to liquid solutions and it was the inability of certain substances to diffuse through membranes such as parchment which, as is well known, led Thomas Graham in 1861 to call such substances *colloids* (from the Greek, $\kappa o \lambda \lambda \alpha = glue$, because they were often sticky and gelatinous as hydrates), and to distinguish them from the smaller dissolved species, which he called *crystalloids*. The latter term is now defunct but the term colloid may be used to include both inorganic colloids and biological macromolecules, which were later classed as hydrophilic colloids because of their affinity for water. The inorganic colloids are mainly physical aggregates of smaller particles and the concept of a macromolecule was only later developed, in the 1920s, and applied to proteins, polysaccharides and other colloids of biological origin. The observation of Graham not only served to delineate a class of substances of great interest in biology but also encouraged the use of *dialysis*, the separation of crystalloids from colloids by means of membranes. Thus osmotic pressure came to be distinguished as that pressure which must be applied to the solution of an impermeable solute so that equilibrium is attained and the net flow of diffusible components across a semi-permeable membrane, separating it from more solvent and diffusible components, is zero. Understanding of the phenomenon was put on a sound quantitative basis by the work of Pfeffer and de Vries (1889) and of J. H. van't Hoff (1888) on the osmotic pressure of aqueous solutions of sucrose, studies made possible because it was found that films of copper (II) ferrocyanide deposited in earthenware pots were permeable to water but not to sucrose. They showed that, provided the solutions were very dilute, the observed osmotic pressures π obeyed the relation,

$$\pi = RTC_2$$

where C_2 is the concentration of an impermeable solute (sucrose) in moles per litre of solution. In the early part of this century, these experiments were greatly improved in technique and accuracy by H. N. Morse and J. C. W. Frazer at Johns Hopkins University, Baltimore, and by the Earl of Berkeley and E. G. J. Hartley at Oxford and it was found that the relation

$$\pi = RT.m',$$

where m' = volume molality, the moles of sucrose per litre of solvent, is obeyed to higher concentrations than that of van't Hoff.

The first studies on the osmotic pressures of protein solutions[†] were made by Starling (1899) followed by Reid (1904, 1905). Starling separated the proteins from blood serum by filtration through gelatin membranes and then measured the osmotic pressure needed to be applied hydrostatically to the unfiltered serum to attain equilibrium and no net flow of water when it was separated from the filtered serum by a membrane of calf peritoneum mounted on silver gauze coated with gelatin. These measurements were of fundamental significance in physiology since the osmotic pressure proved to be of an order of magnitude (*ca.* 30 mm Hg) which was intermediate between the blood pressures in the arteries and the veins, so that in the former water and salts would flow out of the blood and into it in the latter. These earlier studies on the osmotic pressure of proteins and other biological macromolecules, which went *pari passu* with the development of the thermodynamic understanding of dilute solutions (Gibbs 1876), have been admirably surveyed by Adair (1961). Particularly noteworthy were the confirmation by Donnan (1911) of the prediction of Gibbs that diffusible ions would distribute themselves unequally across a membrane impermeable to a charged macro-ion, the experiments of Sørenson (1917) on carefully constituted solutions of egg albumin, the studies of Loeb (1922) on gelatin solutions and the beautifully exact work of Adair himself since 1924, particularly his important value of 66 000 for the molecular weight of haemoglobin (Adair 1925, 1928) which was subsequently confirmed by some of Svedberg's earliest experiments with the ultracentrifuge. These and other osmotic pressure studies have been surveyed in a number of reviews (Gutfreund 1950; Kupke 1960; Adair 1961; Bonner, Dimbat and Stross (mainly synthetic polymers) 1958; Weissberger 1959; Thain 1967) and in more general works (Frazer 1931; Cohn and Edsall 1943; Glasstone 1948; Flory 1953; Tompa 1956; Tanford 1961; Morawetz 1965).

[†] The phrase 'osmotic pressure of a solution' is somewhat loose and ill-defined and really refers to the pressure to be applied to that solution, presumed to contain a macromolecule, to attain equilibrium with respect to movement of the *solvent* when it is separated by a semi-permeable membrane from the solvent (and other diffusible species). It is not a pressure produced *by* the solution, strictly speaking, at all.

After occupying a key position up to the 1940s in the development of an understanding of solutions of biological macromolecules and their physiological roles, osmotic pressure studies have been published more sporadically. A notable contribution was however made by Scatchard and his colleagues (1946, 1954) both in formulating the theory of and making measurements on the osmotic pressure of proteins in solutions of varying ionic constitution. This study involved a careful application of the thermodynamic ideas of Gibbs and of Donnan and this thermodynamic approach has had a significance far wider than stems simply from its relevance to osmotic pressure studies as such. For it is a historical fact that the two phenomena whose study was determinative in establishing that biological 'colloids' were in fact macromolecules involved their solutions under conditions of equilibrium, the one osmotic, the other gravitational in the ultracentrifuge. For this reason, the powerful ideas developed to describe the thermodynamics of solutions not only afford the most rigorous theoretical formulation of the relationships between observations in osmotic and ultracentrifuge studies but also the underlying connections between these two kinds of observation and others, such as light scattering. In this chapter the necessary thermodynamic formulations appropriate to solutions of biological macromolecules are developed to provide the basis for the theory of their osmotic pressure, which is subsequently described. The notation employed is based largely on that of Scatchard *et al.* (1946) and of Casassa and Eisenberg (1960, 1964).

1.1. Definitions and thermodynamic relationships

Thermodynamic components of a mixture will be denoted by $1, 2, \ldots J, K, \ldots$. Since we are primarily concerned with solution properties of biological macromolecules, usually:

1 = solvent (water in most cases);
2 = macro-ion, with or without other small ions;
3 = low molecular weight component, 'diffusible' through membranes impermeable to the macromolecule; usually a simple uni–univalent salt.

Thermodynamic components are almost always electrically neutral even if constituted of two or more molecular or ionic species; otherwise an extra electro-neutrality condition has to be written explicitly along with the specifically thermodynamic conditions defining equilibrium. Moreover, in practice, only such neutral components can be added to a system and the thermodynamic properties of charged components cannot be determined.

Molecular and ionic *species* will be denoted by $1, 2 \ldots i, j \ldots$ and v_{Ji} will represent the number of moles of species i in one mole of component J. When (2) represents a macromolecular component it will be so defined that one mole of (2) contains one mole of the macro-ion, so v_{2i} is the number of molecules or ions of the ith type associated with each mole of macro-ion in one mole of (2).

Concentration scales. In all cases of interest here, one liquid component is in great excess, and is called the solvent. This will be denoted as component (1). The relative amounts of all other components and species can be expressed on various concentration scales, which include the following:

m_J = moles of J/1000 g of solvent (1); the molal scale.
m_i = moles of i/1000 g of solvent (1)
$$= \sum_J v_{Ji} m_J;$$ (1.1)
C_J = moles of J/1000 ml of solution; the molar scale.
c_J = g of J/ml of solution;
w_J = g of J/g of solvent.

Various volume quantities enter into these discussions, in addition to the partial molar and specific volumes. The most important are:

V_m = Volume (in ml) of solution containing 1000 g of solvent (1);
V_m^0 = Value of V_m at infinite dilution of macro-molecular component (2)
   = Volume (in ml) of 1000 g of pure solvent (1);
$\bar{V}_1^0$ = Volume (in ml) of one mole of pure solvent (1)
   (= Partial molar volume of pure solvent, see below). (1.2)

For a two-component system, the mole fraction of J is

$$x_J = \frac{m_J}{1000/M_1 + m_J},$$

where M_1 = molecular weight of solvent (1).

If the solution is dilute, and $x_J \ll 1$, and $m_J \ll 1000/M_1$

$$x_J \simeq \frac{M_1 m_J}{1000}.$$ (1.3)

If 1000 ml of solution has the same volume as 1000 ml of pure solvent

$$x_J \simeq \frac{C_J \bar{V}_1^0}{1000} \quad \text{and} \quad x_J \simeq \frac{c_J \bar{V}_1^0}{M_J},$$ (1.4)

where M_J = molecular weight of J. Generally,

$$C_J = 1000 m_J/V_m; \qquad w_J = m_J M_J/1000;$$
$$c_J = 1000 w_J/V_m = C_J M_J/1000; \qquad m_J/V_m = c_J/M_J;$$ (1.5)

and $V_m/1000$ = volume of solution (in ml) per g of solvent (1).

Partial molar values of extensive thermodynamic properties are defined as follows.

The partial molar value of X of component J is

$$\bar{X}_J = \left(\frac{\partial x}{\partial n_J}\right)_{T,P,n_K},$$

where n_J = number of moles of J

n_K = number of moles of K

$J \neq K$, and X = an extensive thermodynamic property.

The most important case for our purposes is when X is the Gibbs free energy, $G(= U+PV-TS)$, and the partial molal free energy, $\bar{G}_J$, is then also called the *chemical potential* and denoted by

$$\bar{G}_J = \mu_J = \left(\frac{\partial G}{\partial n_J}\right)_{T,P,n_K}, \qquad (1.6)$$

with the same notation as before. Hence, from the basic formulations concerning G,

$$dG = -S.dT + V.dP + \sum_J \mu_J.dn_J \qquad (1.7)$$

at equilibrium.

The chemical potential may also be defined in terms of other thermodynamic functions, for example the Helmholtz free energy, $F(= U-TS)$, by

$$\mu_J = \left(\frac{\partial F}{\partial n_J}\right)_{T,V,n_K} \qquad (1.8)$$

and, at equilibrium, by

$$dF = -S.dT - P.dV + \sum_J \mu_J.dn_J. \qquad (1.9)$$

Because the chemical potential, temperature and pressure are all intensive thermodynamic properties, not dependent on the size of the system, integration leads to the following expressions for the total free energies at a given temperature and pressure:

$$G = \sum_J n_J \mu_J \qquad (1.10)$$

and

$$F = -PV + \sum_J n_J \mu_J.$$

The relationship between these intensive variables is obtained by comparing the complete differentials of the equations (1.10) with the earlier equations

6 THERMODYNAMICS OF MACROMOLECULAR SOLUTIONS

(1.7) or (1.9), and this yields the Gibbs–Duhem equation:

$$-S\,dT + V\,dP - \sum_J n_J\,d\mu_J = 0 \tag{1.11}$$

or

$$\sum_J n_J\,d\mu_J = V\,dP - S\,dT$$

$$= V\,dP, \quad \text{at constant } T$$

$$= 0, \quad \text{at constant } T \text{ and } P.$$

Partial thermodynamic quantities can be related to each other in the same way as are the parent quantities.

e.g.,

$$\mu_J = \bar{H}_J - T\bar{S}_J,$$

$$\left(\frac{\partial(\mu_J/T)}{\partial T}\right)_P = -\frac{\bar{H}_J}{T^2}, \tag{1.12}$$

and

$$(\partial \mu_J/\partial P)_T = \bar{V}_J.$$

Partial specific quantities are the partial quantities obtained when the differentiations are made with respect to the weight (in g) of a particular component J. In particular, the partial specific volume is

$$\bar{v}_J = \left(\frac{\partial(V_m/1000)}{\partial w_J}\right)_{T,P,m}.$$

Conditions of equilibrium. In a system at equilibrium at constant temperature and pressure, the Gibbs free energy is at a minimum so for any small displacement of the system $dG = 0$; and, at constant temperature and volume, correspondingly $dF = 0$.

When components $1, 2 \ldots J \ldots$ are present in phases $\alpha, \beta \ldots$, then if equilibrium is established with respect to a particular component, J:

$$\mu_J^\alpha = \mu_J^\beta = \ldots,$$

where superscripts denote the phases in which J is present.

1.2. Ideal solutions

Chemical potential

An 'ideal' mixture or solution is *defined* as one for which the chemical potential of a component J is related to its mole fraction in the mixture by

$$\mu_J = \mu_J^0 + RT \ln x_J \tag{1.13a}$$

and

$$d\mu_J = RT\,d\ln x_J,$$

where R is the gas constant and μ_J^0 is a standard chemical potential dependent

THERMODYNAMICS OF MACROMOLECULAR SOLUTIONS 7

only on the temperature and pressure. All the laws of ideal mixtures can be derived from this relationship.

Mole fractions are not the most convenient units for use when solutions are under consideration for the solvent component (1) is usually in excess of any other component. The defining relation (1.13a) then becomes:

$$\mu_J = \mu_J^0 + RT \ln m_J, \tag{1.13b}$$

$$\mu_J = \mu_J^0 + RT \ln C_J, \tag{1.13c}$$

or
$$\mu_J = \mu_J^0 + RT \ln c_J, \tag{1.13d}$$

where the μ_J^0 in equations (1.13b, c, d) differ from the μ_J^0 in (1.13a) by $(RT \ln M_1/1000)$, $(RT \ln \bar{V}_1^0/1000)$, and $(RT \ln \bar{V}_1^0/M_J)$, respectively.

In the particular case of a 2-component solution (1 = solvent, 2 = solute), we have

$$\mu_1 - \mu_1^0 = RT \ln x_1 = RT \ln(1-x_2);$$

and, if the solution is not only ideal, but also *dilute*, so that $x_2 \ll 1$, then

$$\ln(1-x_2) = -x_2 - \tfrac{1}{2}x_2^2 \ldots,$$

so that
$$\mu_1 - \mu_1^0 = -RT(x_2 + \tfrac{1}{2}x_2^2 + \ldots);$$

and also

$$\mu_1 - \mu_1^0 = -RT\bar{V}_1^0 c_2 \left[\frac{1}{M_2} + \left(\frac{\bar{V}_1^0}{2M_2^2} \right) c_2 + \ldots \right], \tag{1.14a}$$

since (1.4) then applies. Notice that, if no other consideration were involved, the second term inside the square bracket should become vanishingly small for macromolecules.

This last equation (1.14a) is the basis of the determination of the molecular weights of solute molecules in ideal, dilute solutions by means of the so-called colligative† properties of freezing point depression, elevation of boiling point and osmotic pressure. These properties are all primarily those of the solvent (1), and are related to the solute through eqn 1.14a which expresses the dependence of the chemical potential of the solvent on the concentration of the solute (2). The second term in the brackets on the right hand side, that involving $(\bar{V}_1^0/2M_2^2)$ is normally extremely small compared with the first, especially if the solute is a macromolecule. It may then be ignored and the equation employed in the form:

$$\mu_1 - \mu_1^0 = -RT\bar{V}_1^0 c_2/M_2. \tag{1.14b}$$

If a mixture is ideal over the whole range of possible compositions, then the μ_J^0 in eqn (1.13a) are the chemical potentials of each of the pure components J when $x_J = 1$.

† Since they depend on the *number* of solute molecules per unit volume, see section **2.2**, p. 30.

2

However such a range of ideality is a rare occurrence and more usually ideal behaviour is only approached in dilute solutions, with solvent component(1) in great excess. Then μ_1^0 is indeed the chemical potential of pure (1), which is the same as the free energy per mole of (1) at the particular temperature and pressure. However $\mu_2^0, \ldots \mu_J^0, \ldots$ for the solutes are no longer the chemical potentials of pure 2, ... J, ... etc., but hypothetical limiting values to which the chemical potentials would tend as $x_J \to 1$, *if* the relationship (1.13a) were to hold at all compositions. Thus for ideal, dilute solutions the constants μ_J^0 while still being independent of temperature and pressure are not independent of the nature of the solvent.

As long as this distinction is borne in mind, the notation μ_J^0 can be used for the constant terms in eqn (1.13) for both solvent and solutes. The constant μ_J^0 for a solute J may be regarded as the chemical potential which pure J would hypothetically possess if each molecule of J had the same free energy in the pure state as it has when entirely surrounded by solvent (1) molecules in an infinitely dilute solution of J in (1). This can be regarded as the hypothetical standard state of J, with reference to its solutions in (1).

Thermodynamic functions for mixing

Ideal mixtures and solutions can be viewed in another way which is often quite helpful when macromolecular solutions are being considered. This approach begins by describing the mixing properties of ideal solutions in terms of quantities ΔX_{mix}, where X is again any extensive thermodynamic property. ΔX_{mix} is called the 'X of mixing' and is given by

$$\Delta X_{\text{mix}} = X \text{ of total mixture} - \sum_J n_J X_J^0, \qquad (1.15)$$
$$\text{(or solution)}$$

where the superscript ^{0} to $\bar{X}_J$ refers to pure J or to J in the hypothetical standard state already mentioned.

For example (since $\mu_J^0 \equiv G_J^0$), the free energy of mixing is

$$\Delta G_{\text{mix}} = G - \sum_J n_J \mu_J^0. \qquad (1.16)$$

The quantity ΔG_{mix} therefore represents the change of free energy when the separated constituents of the mixture (or solution) in their pure or hypothetical standard state and in quantities n_J are mixed to produce the actual mixture (or solution) under consideration.

Since the total free energy G of *any* mixture is given by eqn (1.10)

$$\Delta G_{\text{mix}} = \sum_J n_J (\mu_J - \mu_J^0) \qquad (1.17)$$

and, if the mixture (or solution) is ideal, by eqn (1.13a),

$$\Delta G_{\text{mix}}^{\text{ideal}} = RT \sum_J n_J \ln x_J. \qquad (1.18)$$

The ideal heat of mixing turns out to be zero, by the following argument. In general, for any mixture

$$\Delta H_{mix} = (-T^2)[\partial(\Delta G_{mix}/T)/\partial T]_{P,n_J}.$$

Hence
$$\Delta H_{mix}^{ideal} = (-T^2)\left[\partial\left(R\sum_J n_J \ln x_J\right)/\partial T\right]_{P,n_J}.$$

Since the differentiation is made under conditions of constant composition, n_J and $\ln x_J$ inside the summation are constant and so the differential is zero.

$$\therefore \Delta H_{mix}^{ideal} = 0, \tag{1.19}$$

which means that $\bar{H}_J$ in the mixture is independent of the concentration of J and is always equal to $\bar{H}_J^0$ in the pure or hypothetical standard state. Hence

$$\Delta S_{mix}^{ideal} = \frac{1}{T}(\Delta H_{mix}^{ideal} - \Delta G_{mix}^{ideal})$$

$$= -R\sum_J n_J \cdot \ln x_J, \tag{1.20}$$

and
$$\Delta V_{mix}^{ideal} = (\partial(\Delta G_{mix}^{ideal})/\partial P)_{T,n_J} = 0, \tag{1.21}$$

which means, as with H_J, that $\bar{V}_J$ of the mixture is independent of the concentration of J and is always equal to $\bar{V}_J^0$ in the pure or hypothetical standard state. Eqns (1.18)–(1.21) can now be regarded as defining ideal solutions; division of them by $\sum_J n_J$ replaces the n_J by x_J and gives the G, etc., of mixing per mole of solution.

A further transformation of the mixing functions is possible. The change of the mixing functions, ΔX_{mix}, with an infinitesimal increase in the amount of component J involved in the mixing process at constant temperature and pressure is called the 'partial molar X of dilution of J' ($\overline{\Delta X_J}$) and is given by

$$\overline{\Delta X_J} = \left(\frac{\partial(\Delta X_{mix})}{\partial n_J}\right)_{T,P,n_K} = \left(\frac{\partial X}{\partial n_J}\right)_{T,P,n_K} - \frac{\partial}{\partial n_J}\sum_J n_J X_J^0$$

$$= \bar{X}_J - \bar{X}_J^0, \tag{1.22a}$$

from the definition (1.15) for any mixture. In particular,

$$\overline{\Delta G_J} = \overline{\Delta H_J} - T\overline{\Delta S_J}, \tag{1.22b}$$

10   THERMODYNAMICS OF MACROMOLECULAR SOLUTIONS

For an ideal mixture (or solution), these partial molar quantities assume the simple forms (on the basis of eqn 1.18 to 22)

$$\overline{\Delta G_J^{\text{ideal}}} = \mu_J - \mu_J^0 = RT \ln x_J;$$
$$\overline{\Delta H_J^{\text{ideal}}} = 0;$$
$$\overline{\Delta S_J^{\text{ideal}}} = -R \ln x_J;$$
and
$$\overline{\Delta V_J^{\text{ideal}}} = 0.$$
(1.23)

This set of relationships (1.23) may also be used to define an ideal mixture or solution, since they are based on eqns (1.18)–(1.21) which arise from the definition $\mu_J = \mu_J^0 + RT \ln x_J$ (eqn 1.13a). The set of relationships (1.23) has the special virtue as a definition of the ideal state that they can be related to a statistical molecular model of mixing by making use of the usual statistical-thermodynamic relation that $S = k \ln W$ or $\Delta S = k\Delta(\ln W)$, where W is the number of complexions of an assembly. It can be shown (e.g. Caldin 1958, p. 369 ff.) that if two types of molecules A and B are mixed randomly without change of internal energy or heat content (i.e. $\Delta H_{\text{mix}} = 0$) and if molecules A and B are sufficiently alike in size and shape to be able to mix randomly, or, more precisely, to pack in the same way when mixed as they pack in the pure liquids, then

$$\Delta S_{\text{mix}} = -R(x_A \ln x_A + x_B \ln x_B),$$

which is simply eqn (1.20) for two components. This is therefore the molecular and statistical basis of the equations for the ideal entropies and free energies of mixing (since $\Delta G_{\text{mix}}^{\text{ideal}} = \Delta H_{\text{mix}}^{\text{ideal}} - T \cdot \Delta S_{\text{mix}}^{\text{ideal}}$). As will transpire, deviations from ideality can therefore be expressed either as deviations from the condition

$$\mu_J = \mu_J^0 + RT \ln x_J \quad (1.13a)$$

or from the conditions of the expressions for $\Delta X_{\text{mix}}^{\text{ideal}}$, (1.18 to 21) and for $\Delta \overline{X}_{\text{mix}}^{\text{ideal}}$, (1.23).

1.3. Real, non-ideal, solutions

Activity, virial, and osmotic coefficients

The chemical potential of a component of a real solution may be expressed as

$$\mu_J = \mu_J^0 + RT \ln a_J, \quad (1.24)$$

where a_J is the *activity* of component J. The activity may be regarded as a product of a concentration and an *activity coefficient* that differs according to the concentration scale employed. Thus the chemical potential may be written:

$$\mu_J = \mu_J^0 + RT \ln x_J f_J; \quad (1.25)$$
$$\mu_J = \mu_J^0 + RT \ln m_J \gamma_J; \quad (1.26a)$$
or
$$\mu_J = \mu_J^0 + RT \ln m_J + RT \beta, \text{ where } \beta_J = \ln \gamma_J; \quad (1.26b)$$

and
$$\mu_J = \mu_J^0 + RT \ln c_J y_J \quad (1.27)\dagger$$

Here f_J, γ_J and y_J are all activity coefficients on the different concentration scales. The μ_J^0 in the equations for the different concentration scales differ from the μ_J^0 in eqn (1.25) by the same factors as before (see text following eqns (1.13)).

As is amply discussed in works on thermodynamics, the definition of activity, and so of activity coefficient, of the components of a solution follows different conventions for the solvent and the solutes since the limiting ideality is only approached at infinite dilution of the solutes as $x_1 \to 1$. The convention normally adopted in practice is that for solvent (1)

$$\mu_1 = \mu_1^0 + RT \ln x_1 f_1 \quad \text{and} \quad f_1 \to 1 \quad \text{as} \quad x_1 \to 1, \quad (1.28a)$$

so that μ_1^0 is the free energy per mole of pure solvent (1); and for solutes,

$$\mu_J = \mu_J^0 + RT \ln m_J \gamma_J \quad \text{and} \quad \gamma_J \to 1 \quad \text{and} \quad \ln \gamma_J \to 0 \quad \text{as} \quad m_J \to 0 \quad (1.28b)$$

or
$$\mu_J = \mu_J^0 + RT \ln c_J y_J \quad \text{and} \quad y_J \to 1 \quad \text{as} \quad c_J \to 0.$$

The μ_J^0, as with ideal solutions, are not the free energies per mole of pure J but the free energies per mole in a hypothetical state, as previously described and are not the same for the different concentration scales. Deviations from ideality in two-component systems can be expressed not only in terms of the deviations of activity coefficients of solutes from unity, but also in terms of a virial series for the chemical potential of the solvent. The virial character of this expansion follows, by statistical thermodynamic arguments, as a result of a close, and somewhat remarkable, analogy between the relation of the chemical potential of a component in a dilute solution to the potential of average force between solute molecules and that between the pressure of an imperfect gas and the potential of average force between the gas molecules (McMillan and Meyer 1945; Hill 1956, 1958, 1960, ch. 19). The significant restriction in this statistical thermodynamic derivation is that there must be only short-range forces between isolated pairs (or triplets, etc), of solute molecules in the solvent. If the solution of a macro-ion contains a sufficient amount of a simple electrolyte so that the potential between isolated macro-ions is approximately proportional to $\exp(-\kappa r)/r$, because of the shielding effect of the ion atmosphere, then the virial expression (e.g. in the form of eqn 1.29) is applicable even to macro-ionic solutions, although interionic coulombic forces are long-range. This virial series may be written by analogy with eqn (1.14a) and with the corresponding gas equations, as

$$\mu_1 - \mu_1^0 = -RT\bar{V}_1^0 c_2 [M_2^{-1} + Bc_2 + Cc_2^2 + \ldots], \quad (1.29)$$

† Some authors use the symbol y for an activity coefficient in conjunction with the molarity scale (C). Since in this book we shall frequently be concerned with solutions of known weight concentration (c) of macromolecules of unknown molecular weight the above notation has been preferred in this context.

where B and C are called the second and third *virial coefficients*, respectively. The units of B are mol ml g^{-2} and of C are mol ml^2 g^{-3}. The virial equation (1.29) is quite general for non-electrolytes, and so for uncharged macromolecules, and also for charged macromolecules in solutions of the type: (1) solvent, (2) charged macromolecules, and (3) salt at a higher concentration than (2). Otherwise for electrolytes at low concentrations a $c_2^{\frac{1}{2}}$ term has to be included.

As we shall see, the relationship (1.29) can be used directly in formulating expressions for the osmotic pressure† of non-ideal two-component systems. This virial formulation of the non-ideality of a two-component system is related to the activity coefficient formulation by the relation

$$\ln \gamma_2 = 2BM_2c_2 + \tfrac{3}{2}CM_2c_2^2, \qquad (1.30a)$$

as may be derived by comparing $(\partial \mu_2/\partial c_2)_{T,P}$ obtained in the two formulations (via the Gibbs–Duhem relation in the derivation from eqn (1.29)). Eqn (1.30a) is, by some authors, written as

$$\ln \gamma_2 = \alpha_1 c_2 + \alpha_2 c_2^2, \qquad (1.30b)$$

when $\alpha_1 = 2BM_2$ and $\alpha_2 = \tfrac{3}{2}CM_2$, and by others as

$$\ln \gamma_2 = \mathbf{B}M_2c_2 + \mathbf{C}M_2c_2^2, \qquad (1.30c)$$

when $\mathbf{B} = 2B$ and $\mathbf{C} = 3C/2$. Other forms, summarized in Table 1.1, also occur. In what follows B and C of eqn (1.29) and (1.30a) will be used.

TABLE 1.1
Alternative pairs of symbols for the second and third virial coefficients

Second B	Third C
$\alpha_1/2$	$2\alpha_2/3$
$\alpha_1/2M_2$	$2\alpha_2/3M_2$
A_2	A_3
A_2/RT	A_3/RT
Γ_2/M_2	Γ_3/M_2
B'/M_2	C'/M_2

(Units of B': reciprocal concentration units, i.e., ml. g^{-1} or dl. g^{-1}, according to the concentration scale used.)

† Eqn (1.29) also provides the basis of the equations for the scattering of light by solutions of biological macromolecules and for their behaviour at sedimentation equilibrium (See Appendix). These phenomena, of osmotic pressure, light scattering and sedimentation equilibrium, are thereby all fundamentally related and, inevitably, all include the virial coefficients from eqn (1.29) in the formulations which are used, in practice, to derive the molecular weights, M_2, from all three kinds of experimental study.

An alternative function used in formulating deviations from ideality, especially for electrolyte solutions, is the *osmotic coefficient of the solvent* ϕ as defined by the relation

$$\mu_1 = \mu_1^0 + \phi RT \ln x_1, \qquad \boxed{T, P \text{ constant}} \qquad (1.31a)$$

where $\phi \to 1$ as $x_1 \to 1$ and, of course, $\phi = 1$ for ideal solutions, or

$$\mu_1 - \mu_1^0 = -\frac{RTM_1\phi}{1000} \sum_{i \neq 1} m_i. \qquad (1.31b)$$

The osmotic coefficient is simply a logarithmic function of the activity coefficient since
$$\mu_1 = \mu_1^0 + RT \ln x_1 f_1 = \mu_1^0 + \phi RT \ln x,$$
and so
$$\ln f_1 = (\phi - 1)\ln x_1.$$

Because of this relation, the osmotic coefficient of the solvent differs more from unity than does its activity coefficient and so is often more convenient to use in connexion with dilute electrolyte solutes (for which γ_1 may be as close to unity as 0·9999). The appropriateness of the adjective 'osmotic' for this coefficient will be more obvious later (Section 2.3, p. 34).

Non-ideality and thermodynamic functions for mixing.

The condition for ideality of a solution may be expressed by means of relations between composition and the free energies etc. of mixing, as in eqn (1.18 to 1.21), and between composition and partial molar free energies etc. of mixing, as in eqn (1.23). The molecular and statistical basis of these relations is that ideal mixtures and solutions are those in which the exchange of a molecule of component A, near to a molecule of component B, by a molecule of B, causes no change of energy (H) and the mixing is random, or, at least, the same arrangements occur, statistically speaking, in the mixture as in pure A and pure B. If solutions containing more than two components are under consideration, then the above remarks on the molecular, statistical basis of ideality apply only to pairs of components of which one is the solvent. For to say that a solution containing two or more solutes is ideal does not mean that the two solutes by themselves would form ideal mixtures in the absence of solvent, even if they are liquids. Non-ideality in mixtures or solutions can be regarded as deviations of the thermodynamic functions for mixing from these ideal values. Solutions of uncharged macromolecules usually have very small heats of mixing (ΔH_{mix} and $\overline{\Delta H_J}$), since the monomer units are often similar in size and shape to those of the molecules of typical solvents for macromolecules. However the entropy of mixing is far from ideal and is greater (more positive) when a given number of macromolecules (n_2) are mixed with a given number of solvent molecules (n_1) than if the

14 THERMODYNAMICS OF MACROMOLECULAR SOLUTIONS

same number (n_2) of monomer molecules are mixed ideally with the n_1 solvent molecules. Each macromolecule has an extra positive configurational entropy in the mixture over and above its usual translational (etc.) entropy, since it possesses a large number of possible configurations of equal energy; this is responsible for the entropy of formation of the mixture (the entropy of mixing) being larger than in the ideal case. But, even if the macromolecules are not flexible, there is an extra positive contribution to the entropy of mixing on account merely of their greater volume per molecule. This volume is often called the 'excluded volume' but includes other than simple size effects: for example, attractive forces between the macromolecules reduces their apparent 'excluded volume'. However, if the comparison is made on a volume, that is lattice-site, basis then the entropy of mixing of n_2 macromolecules with n_1 solvent molecules is less than the ideal entropy of mixing of vn_2 monomer molecules with n_1 solvent molecules, where v is the number of monomer segments in each macromolecule. From either viewpoint the entropy (and partial molar entropy) of mixing have non-ideal values.

This non-ideal entropy of mixing is chiefly responsible for the non-ideality of the osmotic pressure relationships, for example, of solutions of neutral macromolecules. When the macromolecules are charged, non-ideality also arises from the electrostatic interactions, both attractive and repulsive, between the macromolecular ion and other such ions and small ions which are nearby. In addition, there is the further requirement that the solution of such charged macromolecules must be electrically neutral and this imposes restrictions on the number and kind of small ions which can be present and poses problems about the definition of thermodynamic components for such systems. In order to deal with such problems some further developments of notation are necessary.

1.4. Notation of Scatchard (1946): charged macromolecules

It is convenient to write, as before, that

$$\ln \gamma_J = \beta_J \quad (1.26b)$$

and in addition that

$$\left(\frac{\partial \ln \gamma_J}{\partial m_K}\right)_{T,P,n_1,m} = \left(\frac{\partial \beta_J}{\partial m_K}\right)_{T,P,n_1,m} = \beta_{JK}; \quad (1.32)$$

and

$$\left(\frac{\partial \ln a_J}{\partial m_K}\right)_{T,P,n_1,m} = a_{JK} = \left(\frac{\partial \ln a_K}{\partial m_J}\right)_{T,P,n_1,m} = a_{KJ}, \quad (1.33)$$

since

$$\frac{1}{RT} \cdot \frac{\partial}{\partial m_K}\left(\frac{\partial G}{\partial m_J}\right)_{T,P,n_1,m} = \frac{1}{RT} \cdot \frac{\partial}{\partial m_J}\left(\frac{\partial G}{\partial m_K}\right)_{T,P,n_1,m}.$$

In eqns (1.32) and (1.33) the subscripts mean that the differentiations are made, respectively, under conditions of constant temperature, pressure, number (n_1) of moles of solvent and constant molality of all components other than that with respect to which differentiation is made. These lengthy subscripts will normally be omitted.

For a *two component system* in which component 2 is *neutral*

$$a_2 = m_2\gamma_2$$

and

$$a_{22} = \frac{1}{m_2} + \frac{\partial \ln \gamma_2}{\partial m_2} = \frac{1}{m_2}(1+\beta_{22}m_2). \tag{1.34}$$

From eqn (1.31a) relating activity and virial coefficients and the proportionality between m_2 and c_2 in dilute solutions (so that $\partial \ln m_2/\partial \ln c_2 = 1$), it follows that

$$a_{22} = \frac{1}{m_2}(1+2BM_2c_2+3CM_2c_2^2+\ldots). \tag{1.35}$$

which shows the relation between the Scatchard and virial coefficient notations. It is as a_{22}, expressed as eqn (1.35), that deviations from non-ideality enter the equations for osmotic pressure as well as for sedimentation equilibrium and light scattering. (The apparently curious mixture of molality and weight-concentration terms in (1.35) ultimately proves convenient, since the 'm_2' disappears in equations of practical application.)

For a *two-component system* in which component 2 is *charged*, each mole of 2 contains one mole of macro-ion P and z_2 moles of univalent counterion (X) of total molality z_2m_2. The activity of 2 (i.e. of 1 mole of PX_{z_2}) can then be written as

$$a_2 = m_2(x_2m_2)^{z_2}\gamma_2,$$

whence

$$a_{22} = \frac{1}{m_2} + \frac{z_2}{m_2} + \beta_{22} = \frac{z_2+1}{m_2}\left(1+\frac{\beta_{22}m_2}{z_2+1}\right) \tag{1.34a}$$

or

$$a_{22} = \frac{z_2+1}{m_2}(1+2BM_2c_2+\ldots) \tag{1.35a}$$

The last equality is written by analogy with (3.38). Note that in this instance, when the macromolecule is charged and no other salt is present, any extrapolation to $c_2 = 0$ would still leave the factor (z_2+1) on the right hand side.

Consider a *three-component system* containing the following components: (1) solvent; (2) macromolecule, 1 mole of which contains 1 mole of macroion and ν_{2i} moles of the various diffusible species i; and (3) salt, made up of small ions, i. Then

$$\ln a_2 = \ln(m_2\ldots(m_i)^{\nu_{2i}}\ldots) + \ln \gamma_2$$
$$= \ln m_2 + \sum_i \nu_{2i} \ln m_i + \beta_2, \tag{1.36}$$

16 THERMODYNAMICS OF MACROMOLECULAR SOLUTIONS

where $m_i = $ *total* molalities of ions i in the solution. Differentiation with respect to m_2 gives

$$a_{22} = \frac{1}{m_2} + \sum_i \frac{\nu_{2i}^2}{m_i} + \beta_{22} \tag{1.37}$$

since

$$\nu_{2i} = dm_i/dm_2.$$

If the small ions consist of one kind of small cation (subscript $+$) and one kind of small anion (subscript $-$) only, and m_+ and m_- are their respective molalities,

$$a_{22} = \frac{1}{m_2} + \left(\frac{\nu_{2+}^2}{m_+} + \frac{\nu_{2-}^2}{m_-}\right) + \beta_{22}. \tag{1.38}$$

The quantity a_{23} is given by

$$a_{23} = \sum_i \nu_{2i}\left(\frac{\partial \ln m_i}{\partial m_i}\right)\left(\frac{\partial m_i}{\partial m_3}\right) + \beta_{23}$$

$$= \sum_i \frac{\nu_{2i}\nu_{3i}}{m_i} + \beta_{23}, \tag{1.39}$$

since

$$\partial \ln m_2/\partial m_3 = 0.$$

Again when only two kinds of small ions are present, this becomes

$$a_{23} = \left(\frac{\nu_{2+}\nu_{3+}}{m_+} + \frac{\nu_{2-}\nu_{3-}}{m_-}\right) + \beta_{23}. \tag{1.40}$$

The quantity a_{33} is given by

$$a_{33} = \sum_i \frac{\nu_{3i}^2}{m_i} + \beta_{33}, \tag{1.41}$$

where ν_{3i} is the number of moles of ions of type i per mole of salt component (3). When only two kinds of small ions are present, this becomes

$$a_{33} = \frac{\nu_{3+}^2}{m_+} + \frac{\nu_{3-}^2}{m_-} + \beta_{33}. \tag{1.42}$$

It is important to note that, quite generally, the so-called 'binding coefficient' Γ_{32} (Williams *et al.* 1958) of (3) on (2) is

$$\Gamma_{32} = \left(\frac{\partial m_3}{\partial m_2}\right)_{\mu_3} = -\frac{a_{23}}{a_{33}}, \tag{1.43}$$

since

$$\left(\frac{\partial \ln a_3}{\partial m_2}\right)_{m_3} \times \left(\frac{\partial m_2}{\partial m_3}\right)_{\mu_3 \text{ constant}}^{\ln a_3} \times \left(\frac{\partial m_3}{\partial \ln a_3}\right)_{m_2} = -1$$

and $a_{23} = a_{32}$.

The left hand side of eqn (1.43) represents the change in molality of m_3 per unit change in molality of m_2 at constant chemical potential of component (3). It therefore represents the effective amount of (3) thermodynamically 'bound' per mole of (2), i.e. the excess or deficit of (3), in a solution in which (3) is at constant chemical potential. This quantity therefore corresponds to the excess or, more usually, to the deficit of (3) per mole of macromolecular component (2), when this three-component solution is in osmotic equilibrium, with respect to (3), with a solution of (3) at constant chemical potential —that is at constant concentration of (3), since (2) is absent from the diffusate.

1.5. Ideal osmotic (Donnan) equilibrium

When a three-component system containing only two types of small diffusible ions (+ and −) is at equilibrium in an osmotic experiment of the kind just described and all activity coefficients are unity, so that the solutions are *ideal* (this is usually called the Donnan case after F. G. Donnan who first analysed it (1911, 1935)), then it can be shown (Overbeek 1956; Williams, Van Holde, and Baldwin 1958; Tanford 1961) that the numbers of diffusible ions 'bound' to each macromolecule, i.e. the excess of each per macromolecule, are:

$$v_{2+} = -v_{2-} = -z_2/2,$$

where z_2 is the algebraic positive charge on each macromolecular ion (i.e. z_2 is numerically negative if the macromolecule is negatively charged). Then the total molalities of cation and anion are:

$$m_+ = m_3 - z_2 m_2/2$$

and
$$m_- = m_3 + z_2 m_2/2; \tag{1.44}$$

and each mole of component (2) may be regarded as comprised of:
1 mole of macromolecule, charge $+z_2$,
$-z_2/2$ moles of cations, say Na^+,
$+z_2/2$ moles of anions, say Cl^-,
and component (3), of molality m_3, which is all the Na^+ and Cl^- not part of (2).

This is the definition of Scatchard *et al.* (1946) of the macromolecular component (2), which is discussed more fully below. Substituting the values for m_+ and m_- given by eqn (1.44) and $v_{2+} = -v_{2-} = -z_2/2$, into the appropriate equations for such a three-component system (1.38, 1.40, 1.42), yields

$$a_{22} = \frac{1}{m_2} + \frac{z_2^2}{2m_3 \varepsilon} + \beta_{22}, \tag{1.45}$$

where
$$\varepsilon = 1 - (z_2 m_2 / 2m_3)^2.$$

Thus

$$a_{22} = \frac{1}{m_2}\left[1 + \frac{z_2^2 m_2}{2m_3} + \beta_{22}m_2\right], \qquad (1.46)$$

if $|z_2 m_2| \ll 2m_3$ and hence $\varepsilon \to 1$. Moreover,

$$a_{23} = -\frac{z_2^2 m_2}{2m_3^2 \varepsilon} + \beta_{23}, \qquad (1.47)$$

and

$$a_{33} = \frac{2}{m_3 \varepsilon} + \beta_{33}. \qquad (1.48)$$

The binding coefficient, Γ_{32}, for this situation is then given by

$$\Gamma_{32} = -\frac{a_{23}}{a_{33}} = -\frac{(-z_2^2 m_2/2m_3 + \beta_{23}m_3\varepsilon)}{2 + \beta_{33}m_3\varepsilon}$$
$$= +z_2^2 m_2/4m_3, \qquad (1.49)$$

if $\beta_{23}m_3\varepsilon$ and $\beta_{33}m_3\varepsilon$ are small compared with the first terms in the numerator and denominator, respectively. Edsall et al. (1950) have shown that this is so, except for certain extreme cases, and that the other approximations in eqn (1.46) and (1.47) are also normally justified.

1.6. The definition of the macromolecular component (2)
The term $m_2(a_{22} - a_{23}^2/a_{33})$.

As we shall see in the next chapter, the basic expression for the osmotic pressure (as well as those for sedimentation equilibrium and light scattering) of three-component systems contains the term $m_2(a_{22} - a_{23}^2/a_{33})$, along with the experimental variables and the molecular weight (M_2) of the macromolecular components. If a 3-component system is at ideal Donnan osmotic equilibrium and contains only two types of small diffusible ions ($+$ and $-$), then substitution of the above expressions (1.45 to 48), based on the Scatchard definition of component 2, for a_{22}, a_{23}, and a_{33}, shows that in this case (if $\varepsilon \to 1$ and $|z_2 m_2| \ll 2m_3$)

$$m_2(a_{22} - a_{23}^2/a_{33}) = 1 + m_2\left(\frac{z_2^2}{2m_3} + \beta_{22} - \frac{\beta_{23}^2 m_3}{2 + \beta_{33}m_3}\right) \qquad (1.50a)$$

$$= 1 + \frac{V_m^0}{M_2}\left(\frac{z_2^2}{2m_3} + \beta_{22} - \frac{\beta_{23}^2 m_3}{2 + \beta_{33}m_3}\right)c_2 \qquad (1.50b)$$

$$= 1 + 2B'M_2 c_2, \qquad (1.50c)$$

by analogy with eqn (1.35). Thus a 'second virial coefficient', B' on a concentration scale, can be attributed to a three-component system such that

$$B' = \frac{V_m^0}{2M_2^2}\left(\frac{z_2^2}{2m_3} + \beta_{22} - \frac{\beta_{23}^2 m_3}{2 + \beta_{33}m_3}\right). \qquad (1.51a)$$

THERMODYNAMICS OF MACROMOLECULAR SOLUTIONS 19

If such a three-component system is at osmotic equilibrium, but not under Donnan conditions (i.e. $v \neq -z_2/2$), then the following general equation is obtained instead of eqns (1.50) by substituting for a_{22} the more general expression (1.37):

$$m_2(a_{22} - a_{23}^2/a_{33}) = 1 + 2B''M_2c_2, \qquad (1.50d)$$

where

$$B'' = \frac{V_m^0}{2M_2^2}\left[\sum_i \frac{v_{2i}^2}{m_i} + \beta_{22} - a_{23}^2/a_{33}\right]. \qquad (1.51b)$$

More detailed expressions are obtainable by further expanding the summation and a_{23} and a_{33} by means of eqns (1.38–1.42).

Osmotic equilibrium and definition of the macromolecular component in a three-component system

It is important for the understanding of recent developments of the theory of three-component solutions, notably those expounded by Casassa and Eisenberg (1960, 1964) which, although at first sight complex, eventually simplify procedures and relationships, that if the macromolecular component (2) could be defined so that

$$a_{23}^* = a_{32}^* = 0 \qquad (1.52)$$

(where asterisks show that such a definition of (2) is operative) then, since $a_{33} \neq 0$,

$$m_2^*(a_{22}^* - a_{23}^{*2}/a_{33}) = m_2^* a_{22}^* = 1 + m_2^*\left[\sum_i \frac{v_{2i}^{*2}}{m_i} + \beta_{22}^*\right] \qquad (1.53a)$$

$$= 1 + 2B^* M_2^* c_2^* + \text{higher terms} \qquad (1.53b)$$

by analogy with eqn (1.35); with

$$B^* = V^0/2M_2^{*2} \times [\text{square bracket in eqn (1.53a)}]. \qquad (1.54)$$

Any definition of (2) which reduces a_{23}^* and a_{32}^* to zero must also mean that

$$(\partial \mu_2^*/\partial m_3)_{m_2^*} = 0 = (\partial \mu_3/\partial m_2^*)_{m_3} \qquad (1.55)$$

so that in defining (2) as (2*) under condition (1.52), one is also ensuring that the chemical potential of (3) does not change on adding this macromolecular component (2*). Furthermore, since $a_{33} \neq 0$, if (2) is defined so that $a_{23}^* = 0 = a_{32}^*$, then

$$\Gamma_{32}^* = -\frac{a_{23}^*}{a_{33}} = \left(\frac{\partial m_3}{\partial m_2^*}\right)_{\mu_3} = 0. \qquad (1.56)$$

This relation gives the clue to the desired definition of the macromolecular component, since from it we see that (2) must be so defined that, when (2*) is added under conditions where the chemical potential (μ_3) of (3) is constant, there is no change in the molality (m_3) of component (3). The simplest experimental conditions exhibiting constancy of μ_3 is that already described for the Donnan equilibrium system, namely equilibrium dialysis across a membrane impermeable to (2), but across which (3) is in osmotic equilibrium with external diffusate containing only solvent and component (3) at constant chemical potential, that is at constant concentration. Thus the component (2*) must contain all the excess (or deficit) of small diffusible ions of (3) which are in the macromolecular solution over and above those in the external diffusate. The definition of 2* which is required therefore accounts to a formulation of the quantities v^*_{2i} with reference to an osmotic equilibrium experiment, in which μ_3 is constant. This definition, which is associated with Casassa and Eisenberg (1960, 1964), is broader than that of Scatchard, in which $v_{2i} = |\frac{1}{2}z_2|$, for two types only of cation and anion, and which applies to the dialysis equilibrium situation only when it is ideal and all activity coefficients are unity.

The Scatchard and Cassassa–Eisenberg definitions of a charged macromolecular component

One cannot let the thermodynamic components be simply identified with ionic species because then one more component appears than if components were neutral entities; and this means that the electroneutrality condition has to be written down as one of the equations controlling the system, in addition to the thermodynamic relationships. It is therefore simpler to work with electrically neutral thermodynamic components and this, in any case, accords with experimental practice since only such neutral entities can actually be added to a mixture.

However, another difficulty arises at once. For suppose each macromolecule is a cation of positive charge z_2, then the obvious neutral entity to choose as one mole of macromolecular component (2) might be thought to be one mole of P^{z_2+} and z_2 moles of gegenion, Cl^- say; that is one mole of (2) is $(P^{z_2+} + z_2\, Cl^-)$. But then adding one mole of (2) also means adding z_2 small ions and so this changes the activities of solvent and of both small and macromolecular ions. The changes so effected would be due almost entirely to these z_2 small ions in one mole of (2) rather than to the one P^{z_2+} cation, since z_2 might be of the order of 10 to 100. Thus any change in the activity of the solvent in, for example, an osmotic pressure experiment would be chiefly due to the addition of Cl^- in component (2) and the resulting molecular weight of component (2) inferred from this effect on the solvent would be the (number) average of the weights of one macromolecular cation and z_2 chloride ions, which is not the desired quantity, the molecular weight of the macromolecular ion alone.

This problem has been avoided by definitions of the charged macromolecular component which have been based on the proposition that, if a third component is present, consisting of small diffusible ions (e.g. NaCl), then

ν moles of NaCl can always be included in each mole of $(P^{z_2+}.\text{Cl}_{z_2})$ to constitute the macromolecular component (2) so that one mole of (2) is then:

$$[P^{z_2+}+z_2\,\text{Cl}^-+\nu(\text{NaCl})].$$

The value for ν is chosen so that the macro-ion is alone effective in modifying the chemical potential of the solvent. Two main choices for ν have been proposed.

I. $\nu = -z_2/2$, so that 1 mole of (2) is:

$$[P^{z_2+}+z_2\,\text{Cl}^- -\tfrac{1}{2}z_2(\text{NaCl})] = [P^{z_2+}-\tfrac{1}{2}z_2\,\text{Na}^+ +\tfrac{1}{2}z_2\,\text{Cl}^-],$$

which is the formulation of Scatchard et al. (1946: see also Edsall et al. 1950). When one mole of (2) so defined is added to the mixture, there is a net addition of only one mole of macro-ions, so if the effect of one sodium ion on activities is the same as that of one chloride ion (which is tantamount to an insistence that the solutions are dilute enough for their activity coefficients to be identical), only the added macromolecular ion has any effect on the activities of solvent and of the ions already present, both small and macromolecular. This definition is that already employed in Section (1.5), and, as there stated, the quantity $(-z_2/2)$ moles NaCl is the imbalance per mole of macromolecule (there is less NaCl on the macromolecular side) in an osmotic equilibrium experiment, if all activity coefficients are unity and the protein concentration is tending to zero, which is the classical ideal Donnan equilibrium.

The situation envisaged (Fig. 1.1) in the simplest Donnan form of the osmotic equilibrium experiment is an 'inside' solution containing solvent, charged macromolecular ion and the ions ($+$ and $-$) of a salt separated by a membrane permeable only to solvent and to salt from an 'outside' solution containing solvent and salt, but no macromolecule. Because the solute concentrations are not the same on each side of the membrane, the chemical potentials of the solvent on the two sides also differ and can only become the same if a pressure, the osmotic pressure,† is applied to the inside solution, since the solvent is in great excess on both sides and its mole fraction can adjust itself by diffusion to only a negligible extent. However, the chemical potentials of the diffusible salt on each side of the membrane can adjust themselves to equality by diffusion so that, if activity coefficients are all unity,

$$m_+^{in}m_-^{in} = m_+^{out}m_-^{out},$$

where the m's refer to total concentrations of $+$ and $-$ ions *in*side or *out*side. If the macromolecular concentration inside is m_P and the charge on each

---

† It should be noted that, in the usual type of dialysis experiment, if equilibrium is allowed to be attained, this compression results from the elastic distension of the dialysis bag.

macromolecular ion is z_2 positive charges, then the conditions of electroneutrality are

$$z_2 m_P + m_+^{in} - m_-^{in} = 0 \quad \text{and} \quad m_+^{out} = m_-^{out} \equiv m_\pm^{out}$$

where $m_\pm^{out}$ is the molality of diffusible salt in the outside solution. From this, the network of interrelations

$$\begin{array}{ccc} m_+^{in} < m_+^{out} & & m_+^{in} < m_-^{in} \\ \wedge & \| & \wedge \quad \vee \\ m_-^{in} > m_-^{out} & \text{or} & m_+^{out} = m_-^{out} \end{array}$$

is directly derivable; more particularly, it can also be shown (Donnan, 1911; Overbeek, 1956; Tanford, 1961) that

$$-(m_+^{in} - m_+^{out}) = (m_-^{in} - m_-^{out}) = \tfrac{1}{2} z_2 . m_P, \tag{1.57}$$

as $m_P \to 0$. So that, at equilibrium under Donnan conditions (activity coefficients equal to unity and macromolecular concentration tending to zero) the concentrations of the various species are as shown in Fig. 1.1.

The *components* in the inside solution can be defined as desired. If z_2 cations are grouped with each macromolecular cation, then we have the definition of (2) already rejected as unsatisfactory, viz. $P^{z_2+} + z_2^{small}_{anions}$. But if we define (2) so that $v = -z_2/2$, then, from Fig. 1.1, component (2) is

$$[P^{z_2+} - \tfrac{1}{2} z_2^{small}_{2 cations} + \tfrac{1}{2} z_2^{small}_{2 anions}]$$

and the remaining concentrations of small cations and small anions which are left on the *inside*, and now constitute component (3), become thereby equal to $m_\pm^{out}$, which is their concentration outside, and is constant, so that μ_3 is constant. Hence addition of macromolecular component (2) of this composition to the inside solution would not affect the concentration of the salt component (3), comprised of small diffusible ions. This remains constant and equal to $m_\pm^{out}$, so that μ_3 is also constant.

Component (2) so defined therefore fulfils the desired conditions that $a_{32} = 0$, with all the simplifying consequences described above. However, this definition is restricted to conditions of ideal Donnan equilibrium only, when activity coefficients are unity and $m_p \to 0$: a more general formulation is therefore required.

II. v^* = *Imbalance, per mole of macromolecular ion, of the diffusible salt in an osmotic equilibrium experiment*, a definition elaborated by Casassa and Eisenberg (1960, 1964). (An asterisk is usually added to all symbols referring to the macromolecular component (2*) so defined.)

The general situation is exactly comparable with the Donnan case and is depicted in Fig. 1.2. One mole of component (2*) includes one mole of macromolecular cation, $P^{z_{2*}+}$, together with its z_{2*} g ions of gegions $(-)$

THERMODYNAMICS OF MACROMOLECULAR SOLUTIONS 23

Molecular species	Inside Macromolecular solution	Semi-permeable membrane	Outside Solution of diffusible components only
Solvent Component (1) (In equilibrium if osmotic pressure π applied to the inside†)	H_2O	⇌	H_2O
Macromolecular cation Part of component (2)	P^{z_2+} Molality $= m_p$		Molality $= 0$
Diffusible ions (+ and −) of uni-univalent salt Partly component (3) and partly (2)	⊕ $m_+^{in} = m_\pm^{out}$ ⊖ $m_-^{in} = m_\pm^{out}$	$\left. \begin{array}{l} -\frac{z_2}{2}m_p \\ -\frac{z_2}{2}m_p + z_2 m_p \end{array} \right\}$ ⊕ ⊖ ⇌ ⊕ ⊖	$m_+^{out} = m_\pm$ $m_-^{out} = m_\pm$ $\bigg\}$ Constant

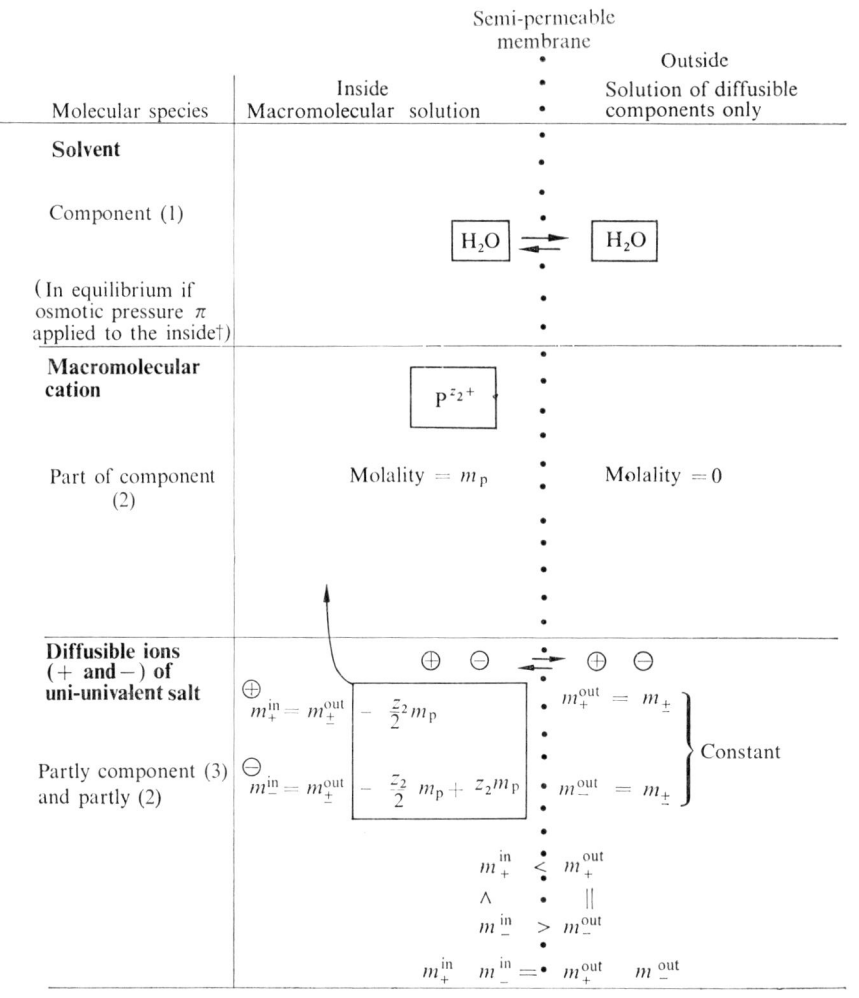

$$m_+^{in} < m_+^{out}$$
$$\wedge \quad \quad \|$$
$$m_-^{in} > m_-^{out}$$
$$m_+^{in} \, m_-^{in} = m_+^{out} \, m_-^{out}$$

FIG. 1.1. Ideal Donnan osmotic equilibrium. The box and arrow represent the combining of these amounts of +ve and −ve ions together with the macromolecular cation to constitute component (2), cf. Scatchard. The dotted line represents a semi-permeable membrane, impermeable to macro-ion. Superscripts 'in' and 'out' refer respectively to the 'inside' macromolecular solution and the 'outside' solution of diffusible components only.

† See footnote to p. 21.

and ν^* moles of neutral salt (+ and −) which may be regarded as the imbalance of neutral salt between the inside and outside solutions: there is usually a deficit on the inside so that ν^* is numerically negative. But when the ions enclosed in the box in Fig. 1.2 are included within the macromolecular component (2*), the remaining concentration of neutral diffusible salt, now component (3), is simply equal to the constant, outside concentration,

24 THERMODYNAMICS OF MACROMOLECULAR SOLUTIONS

Molecular species	Inside Macromolecular solution	Outside Solution of diffusible components only
Solvent Component (1) (In equilibrium if osmotic pressure π applied to the 'inside' †)	H_2O ⇌	H_2O
Macromolecular cation Part of component 2	$P^{z_{2^*+}}$ Molality = m_P	Molality = 0
Diffusible ions (+ and −) of uni-univalent + salt Partly component (3) and partly (2*)	⊕ $m_+^{in} = m_\pm^{out} + v^* m_P$ ⊖ $m_-^{in} = m_\pm^{out} + z_{2^*} m_P + v^* m_P$	⊕ ⊖ ⇌ ⊕ ⊖ $m_+^{out} = m_\pm^{out}$ $m_-^{out} = m_\pm^{out}$ } Constant

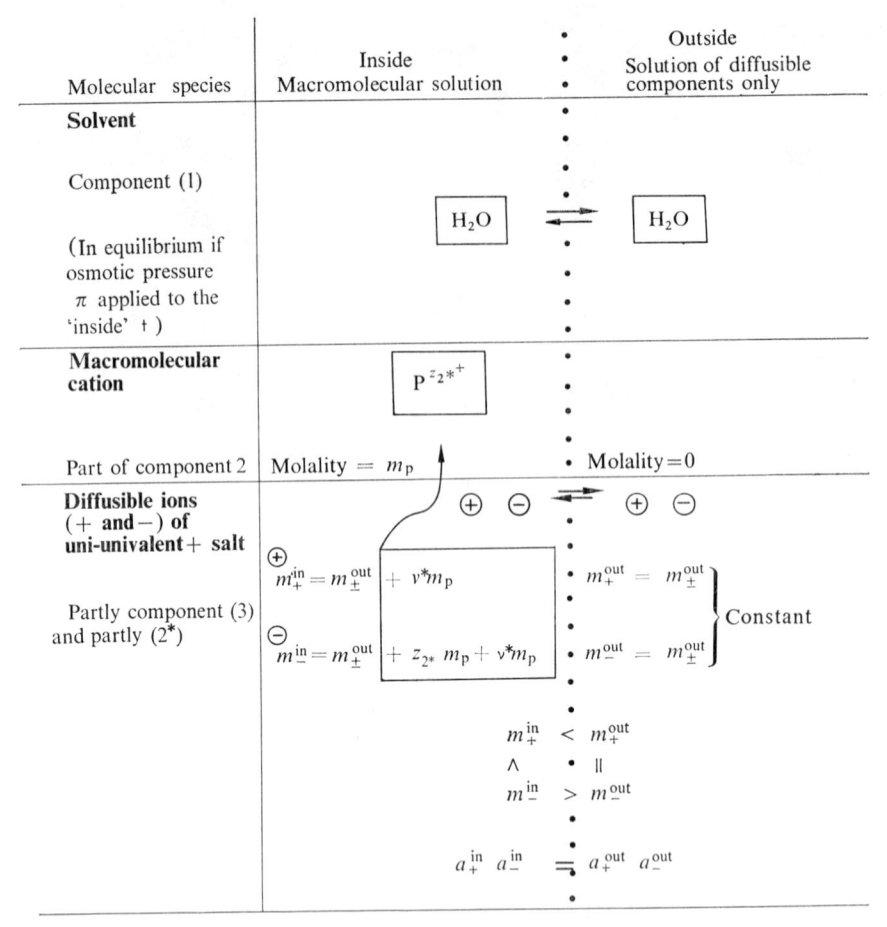

$$m_+^{in} < m_+^{out}$$
$$\wedge \quad \parallel$$
$$m_-^{in} > m_-^{out}$$

$$a_+^{in} a_-^{in} \rightleftharpoons a_+^{out} a_-^{out}$$

FIG. 1.2. Non-ideal osmotic equilibrium. The box and arrow represent the combining of these amounts of +ve and −ve ions together with the macromolecular cation to constitute component (2*), according to the more general definition of Casassa and Eisenberg. The dotted line represents a semi-permeable membrane, impermeable to macro-ion. a = activities. Superscripts 'in' and 'out' as in Fig. 1.1.

† See footnote to p. 21.

$m_\pm^{out}$. So the condition is fulfilled that the addition of 2* does not change the chemical potential and concentration of component (3) which remains equal to its constant value in the outside solution. Thus $a_{23}^* = 0 = a_{32}^*$ which is the condition from which all the other desired relationships follow, as already described above.

The idea of the osmotic equilibrium actually being set up with an outside solution in equilibrium across a membrane can now be set aside since it has served its purpose of revealing what definition of (2) allows a_{32} to become zero. Nevertheless in many types of study, and not only osmotic pressure determinations, it is often convenient to establish osmotic equilibrium by a dialysis procedure. If there is any experimental quantity, Φ (e.g., refractive index or density), which is a sum of contributions from each of the constituent ionic species, which are themselves equal to the product of their concentration and their contribution per mole (ϕ), and the charge on the species is irrelevant, then

$$\Phi^{in} = \phi_{P^{z_{2*}+}} \cdot m_P + \phi_+(m_{\pm}^{out} + \nu^* m_P) + \phi_-(m_{\pm}^{out} + z_{2*} m_P + \nu^* m_P)$$

and
$$\Phi^{out} = \phi_+ m_{\pm}^{out} + \phi_- \cdot m_{\pm}^{out}$$

Hence
$$\Phi^{in} - \Phi^{out} = [(\phi_{P^{z_{2*}+}} + z_{2*} \phi_-) + \nu^*(\phi_+ + \phi_-)] m_P.$$

The term in the square brackets corresponds exactly to the composition of component 2* and since $m_2^* = m_P$, as 2* always contains only one mole of macromolecular ion, the above equation could be more concisely written as

$$\Phi^{in} - \Phi^{out} = \phi_{2*} \cdot m_2^*. \tag{1.58}$$

Hence, the excess of the measured value of Φ (e.g., refractive index or density) for the inner solution over that for the outer diffusate is a quantity directly proportional to the concentration (m_2^*) of 2*, as defined above. It is therefore often necessary to measure in this way the difference between the refractive index, for example, of a dialysed three-component macromolecular solution and that of its diffusate, in order to obtain a quantity proportional to the concentration of 2*, whose definition is now seen to have this added advantage of accessibility of its concentration to direct measurement.

This definition of 2 as 2* has the advantage that it is not restricted, like the Scatchard definition I, to ideal Donnan conditions of osmotic equilibrium and is therefore quite general. A major detraction from this definition would however exist if the quantity ν^*, which is the imbalance of diffusible salt *per mole* of macromolecular ion, were to vary much with the actual concentration, m_P, of macromolecular ion. Fortunately, in the range of concentrations at which biological macromolecules are usually examined there is reasonable experimental evidence, for example with solutions of serum albumin (Scatchard et al. 1946) that ν_{2i}^* is constant over the working range of concentrations, m_P, used in most osmotic pressure (as well as sedimentation equilibrium and light-scattering studies) so that 2* has a constant composition independent of m_2^*, as desired. However, it should be noted that ν_{2i}^* is not independent of m_i so that 2* does not have a constant composition in a gradient of salt ions in, for example, a density-gradient sedimentation experiment.

It should also be noted that the molal *concentration* of macromolecular component (2) is always equal to that of the macromolecular ion, since one mole of (2), in all definitions, contains one mole of this ion so

$$\frac{m_2^*}{1000} = \frac{w_2^*}{M_2^*}$$

$$= \frac{w_2'(= \text{wt. of macromol. per g of solvent, by analysis})}{M_2(= \text{the mol. wt. of macromol. used as the reference for analysis})} \quad (1.59)$$

However, the activity of (2) *does* depend on how it is defined. Conversely, the activities of components (1) and (3) do not depend on the definition of (2), because μ_1 and μ_3 are regarded as kept constant, but the concentration of (3) does depend on the definition of (2). Here w_2' and M_2 are, respectively, the weight/g of solvent and the molecular weight on any definition of (2) other than that of Casassa–Eisenberg (see above). The relationships (1.59) are important since, as we shall see, experiment, in effect, yields m_2^*, but eqn (1.59) shows that the molecular weight which will be derived will depend on the method of analysis used to obtain w_2^* (or w_2) and so on the reference species for this analysis. For example, if the w_2 of proteins are determined by nitrogen analysis and the g(N)/g (solvent) so obtained is multiplied by a factor α to obtain the g protein/g solvent, then the M_2 derived is that of the macromolecular species a fraction $1/\alpha$ of whose weight is nitrogen (usually the isoionic species). If, however, the dry weight per g solvent in the diffusate ('outside' solution) is subtracted from the dry weight per g solvent in the macromolecular dialysed solution, the resulting difference is truly w_2^* and the experiments would yield M_2^*, which includes the $\sum v_{2i}^*$ moles of i.

Exactly parallel considerations apply if the analysis yields concentrations on the c scale, i.e. g of component (2) per ml of solution, which is c_2. The remarks in the preceding paragraph concerning the entity whose molecular weight, M_2, is determined apply in the same sense. The relationships corresponding to eqn (1.59) which are then relevant are usually:

$$\frac{m_2^*}{V_m^0} = \frac{c_2^*}{M_2^*} = \frac{c_2'}{M_2'}, \quad (1.60)$$

where M_2' has the same meaning as in eqn (1.59) and c_2' is the wt. of macromolecule, per ml of solution, as determined by a suitable analytical method. Some methods, notably osmotic pressure, in effect determine m_2^*/V_m^0 and the first equality shows this is simply the molar concentration of the macromolecular component which is, like the molal concentration, not affected by the definition of component (2) one mole of which contains 1 mole of macro-ion on all definitions.

Appendix

Light scattering and sedimentation equilibrium of solutions of biological macromolecules and the virial equation for the chemical potential of the solvent

The virial equation

$$\mu_1 - \mu_1^0 = -RT\bar{V}_1^0 c_2 [M_2^{-1} + Bc_2 + Cc_2^2 + ...] \qquad (1.29)$$

relates the chemical potential of the solvent to that of the solute concentration, c_2, in a two-component system. In the formulation of the relationships for the *scattering of light* by solutions of biological macromolecules it is the rate of change of the chemical potential of the solvent with solute concentration, namely $(\partial \mu_1/\partial c_2)_{T,P}$, which enters the expressions through fluctuation theory (e.g. see Tanford 1961, ch. 5). Differentiation of (1.29) yields at once the desired expression

$$\left(\frac{\partial \mu_1}{\partial c_2}\right)_{T,P} = -\frac{RT\bar{V}_1^0}{M_2}(1 + 2BM_2 c_2 + 3CM_2 c_2^2 + ...) \qquad (A.1)$$

However, the quantity of interest in *sedimentation equilibrium* studies is the rate of change of the chemical potential of solute (2) with its own concentration c_2; that is $(\partial \mu_2/\partial c_2)_{T,P}$. The required expression for this quantity is given by the virial equation (1.29) and by the Gibbs–Duhem relations and is derived as follows. From the Gibbs–Duhem equation (1.11) for a two-component system at constant temperature and pressure

$$x_1 \, d\mu_1 + x_2 \, d\mu_2 = 0$$

and since $x_1 = 1 - x_2$,

$$\frac{d\mu_2}{dx_2} = -\frac{(1-x_2)}{x_2} \cdot \frac{d\mu_1}{dx_2}$$

If the solution is *dilute*, eqn (1.4) applies so

$$x_2 = (\bar{V}_1^0/M_2)c_2, \quad \text{i.e.} \quad dx_2 = (\bar{V}_1^0/M_2) \, dc_2$$

and

$$(1-x_2) \simeq 1.$$

Substitution in the preceding equation then yields

$$\frac{d\mu_2}{dc_2} = -\frac{M_2}{c_2 \bar{V}_1^0} \cdot \frac{d\mu_1}{dc_2}$$

and hence, from (A.1), with re-introduction of the conditions of constant temperature and pressure,

$$\left(\frac{\partial \mu_2}{\partial c_2}\right)_{T,P} = \frac{RT}{c_2}(1 + 2BM_2 c_2 + 3CM_2 c_2^2 + ...) \qquad (A.2)$$

Eqn (A.2) is then used to derive the equation for sedimentation equilibrium (cf. Tanford 1961, ch. 4. Equations (16.13) and (16.15)).

2
THEORY OF OSMOTIC PRESSURE

2.1. Introduction

IN this chapter, the theory of osmotic pressure will be developed so as to relate actual observations—on the temperature, concentration of solutes, the osmotic pressure and permeability of the membrane—to molecular parameters of the size, shape and interactions of the non-permeable macromolecular solute. The principal conceptual tool in this development, beginning in the next section, will be thermodynamic, as already adumbrated in the previous chapter. However some elementary considerations based on the randomness of molecular motions are worth outlining before engaging in the somewhat more abstract thermodynamic exercise, for the development of an osmotic pressure, like many other properties of solutions, depends ultimately on the effects of these motions, as experienced in bulk properties.

Any two-component solution may be considered as uniform in composition even down to quite small volume elements containing only a few molecules of each component, provided the composition is averaged over a time long compared with the time it takes a molecule to traverse the volume element. This uniformity of composition is a direct consequence of the random motion of the molecules and still prevails, even in structured solvents, provided a time-average is taken.

If two distinct volumes of a solution and its solvent are brought into contact so that exchange of material is possible (e.g., via a membrane or a vapour phase), both will alter spontaneously in such a way that the composition of the system becomes uniform throughout. If, for example, solvent were layered on to a solution, the random relative motion of the molecules causes the solvent and solute molecules to become concomitantly and evenly distributed through the combined volume. The most obvious observable effect would be movement of solute into the pure solvent layer (the process of solute diffusion) but there would also be movement of solvent molecules into the solution layer and *one* diffusion coefficient suffices to describe the combined process. The statistics are well known of this random movement of molecules, which is commonly viewed as net transport down a concentration gradient (Einstein 1905; Zimm 1943; Tanford 1961). Suppose now the solution is separated from solvent by a barrier of some kind through which only solvent can pass: this could be a semi-permeable membrane, or it could be a vapour phase, indeed the two are formally identical. The system will again alter spontaneously towards a state of uniformity of composition but the way in which it can do this is now limited for the only net movement possible is that of solvent into the solution. Solvent flow will continue through the membrane until it has all passed through, achieving once more a uniform composition. This end result can be prevented, and the flow reduced, halted, or even reversed, by applying a pressure to the solution. The flow of solvent may, if the geometrical arrangements are appropriate, itself generate a

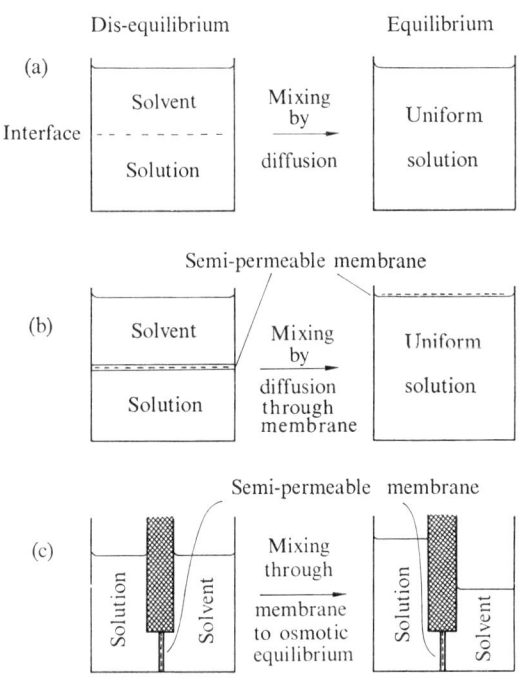

FIG. 2.1. The effect of placing solvent and solution in contact in three different ways. (a) Simple layering, followed by mixing. (b) Through a moveable semi-permeable membrane (c) Through a fixed semi-permeable membrane, with equilibrium established by hydrostatic pressure.

hydrostatic head which is sufficient to stop it; or the solution may be contained in an elastic vessel (e.g. a blood vessel, or a cell membrane or a dialysis sac) which by distortion generates a back pressure. The pressure which is just sufficient to halt the flow of solvent, thus maintaining the system in equilibrium is the osmotic pressure, as already defined. Fig. 2.1 illustrates some common situations.

These considerations show that the osmotic pressure is a consequence of the way solutions tend to adopt a uniform composition, which is itself a consequence of the random motion of the molecules which compose them. We would expect, therefore, that the magnitude of the osmotic pressure will depend on the composition of the solution, and on the absolute temperature.

The tendency of solvent to move into a solution would reasonably be expected to be related to the concentration of solvent in the solution. The lower the concentration of solvent, the greater the tendency to move into it: conversely therefore, the greater the concentration of solute, the greater the osmotic pressure is likely to be. Although we are usually interested chiefly in the solute, osmotic pressures are more easily understood if the solvent is the focus of attention.

It is also quite reasonable to speak of the osmotic pressure of a solution, and to assign values to it even if it is not possible to generate an actual hydrostatic pressure because suitable membranes do not exist. The osmotic pressure is a general bulk property of a particular solution system (i.e., a particular solution and a particular membrane with specified permeabilities), and it can in principle be both defined and measured in pressure units. It is sometimes convenient, instead of the osmotic pressure π, to use π/RT which has the units of mol m^{-3}, and is the same as the former mosmolar unit. At room temperature, if $\pi/RT = 1$ mol m^{-3} then $\pi = 2\cdot5 \times 10^3$ Pa.

2.2. The chemical potential of the solvent in macromolecular solutions and colligative properties

It was shown (Section 1.2) how the presence of any solute inevitably reduced the chemical potential of the solvent (μ_1) to a value below its level in the pure state (μ_1^0) by reducing the mole fraction of the solvent in the solution to below unity. This led to the following relationships, for two-component solutions, between this reduction in the chemical potential of the solvent and the concentration of solute:

$$\mu_1 - \mu_1^0 = -RT\bar{V}_1^0 c_2 \left[\frac{1}{M_2} + \left(\frac{\bar{V}_1^0}{2M_2^2} \right) c_2 + \ldots \right] \tag{1.14a}$$

for dilute, ideal, two-component solutions; and in general,

$$\mu_1 - \mu_1^0 = -RT\bar{V}_1^0 c_2 \left[\frac{1}{M_2} + Bc_2 + Cc_2^2 + \ldots \right] \tag{1.29}$$

for non-ideal two-component solutions. As was pointed out in Section 1.2, when the second term in the square brackets is negligible (i.e. when the solutions are ideal or when $c_2 \to 0$ in non-ideal solutions), both of these equations reduce to

$$\mu_1 - \mu_1^0 = -RT\bar{V}_1^0 c_2 / M_2 \tag{1.14b}$$

Hence any method which affords ($\mu_1 - \mu_1^0$) can also yield an unknown molecular weight M_2, since c_2 and $\bar{V}_1^0$ may be determined by other means. The quantity c_2/M_2 is a molarity which is proportional to the number of molecules of (2) per unit volume and methods which depend on experimental properties determined by ($\mu_1 - \mu_1^0$) are called *colligative* (Latin, *colligatus* = collected together) because such properties depend on the *number* of particles present per unit volume. The reduction in the chemical potential μ_1 of the solvent because of the presence of solute (2) is reflected *inter alia*, in: a lowering of vapour pressure, and so in an elevation of the temperature at which it reaches 760 mm Hg (the boiling point); in a depression of the temperature (the freezing point) at which its chemical potential is equal to that of the pure solid; and in a disequilibrium, since $\mu_1 \neq \mu_1^0$, at a given temperature and pressure if the solution is separated (as in Fig. 2.2) from pure solvent by a membrane permeable only to solvent (1) and not to solute (2). The last-mentioned is, of course, an osmotic experiment and since the solvent is, by definition, in great excess no finite amount of movement of solvent across from the 'outside' into the solution 'inside' can ever increase μ_1 back to its value in the pure state, μ_1^0, since (2) is always present to keep the mole fraction of (1) below unity. Hence at constant temperature and pressure, solvent diffuses into the solution indefinitely, with the decrease of free energy characteristic of a natural, irreversible process. The only way of attaining, at constant temperature, equilibrium with respect to the solvent is by increasing the pressure on the solution side, for $(\partial \mu_J/\partial P)_T = \bar{V}_J$ is always positive and an appropriate increase of pressure π can increase the chemical potential of solvent in the solution up to its value in the pure solvent μ_1^0. As we have stated earlier, the pressure π which does this is called the *osmotic pressure*. It is often loosely described as the 'osmotic pressure of the solution' when strictly it is a pressure affecting a property of the *solvent* in a particular solution system (Section 2.1). Since this property depends on the nature of the solution this terminology is perhaps permissible provided it is always clearly understood that it is the behaviour of the *solvent* which is under consideration. The situation at osmotic equilibrium is depicted schematically in Fig. 2.2. When the extra pressure π applied to the solution side is the osmotic pressure the rate of flow of solvent across the membrane is the same in both directions. Methods of measuring osmotic pressure are all based on this criterion of no net flow whether they seek to achieve this point in practice (static methods) or to extrapolate to it from a position of disequilibrium (dynamic methods). When only two components (1 = solvent, 2 = non-diffusible solute) are present, the equilibrium situation of Fig. 2.2 may be described in thermodynamic terms as follows.

The chemical potential μ_1^{out} of (1) in the outer compartment is equal to the chemical potential of pure solvent, μ_1^0. At constant temperature T, and the same pressure P on both sides of the membrane, the chemical potential μ_1^{in} of the solvent in the solution is less than that outside:

$$(\mu_1^{\text{in}})_P < (\mu_1^0)_P = (\mu_1^{\text{out}})_P, \quad \boxed{\text{Constant } T}$$

where subscripts denote the pressure but not the temperature, since this is constant throughout.

If the pressure on the solvent in the solution 'inside' is increased from P to $(P+\pi)$, the chemical potential of the solvent in the solution is then increased to

$$(\mu_1^{in})_{P+\pi} = (\mu_1^{in})_P + \int_P^{P+\pi} \left(\frac{\partial \mu_1}{\partial P}\right)_T \cdot dP = (\mu_1^{in})_P + \bar{V}_1 \cdot \pi, \quad \boxed{\text{Constant } T} \quad (2.1)$$

provided $\bar{V}_1$, is not a function of pressure, i.e. compressibility can be ignored.

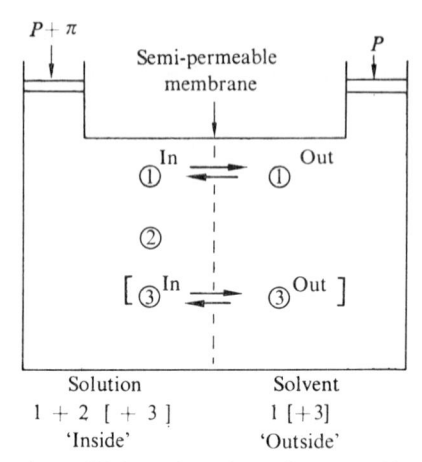

FIG. 2.2. Osmotic equilibrium. (Membrane impermeable to component 2)

If equilibrium is to be achieved this new value, $(\mu_1^{in})_{P+\pi}$, of the chemical potential of the solvent in the solution inside at pressure $(P+\pi)$ must now equal its chemical potential 'outside', $(\mu_1^0)_P$ in the pure solvent still under a pressure P. Hence:

$$(\mu_1^{in})_{P+\pi} = (\mu_1^{out})_P = (\mu_1^0)_P; \quad (2.2)$$

and, when eqns (2.1) and (2.2) are combined,

$$(\mu_1^{in})_P - (\mu_1^0)_P = -\pi \bar{V}_1$$

or, more simply,

$$\mu_1 - \mu_1^0 = -\pi \bar{V}_1, \quad \boxed{\text{Constant } T, P} \quad (2.3)$$

where μ_1 refers to the solution, and all quantities are written for the original pressure P. Eqn (2.3) shows that the quantity $(\mu_1 - \mu_1^0)$, which is related to M_2 and c_2 by the virial equations given above (1.14a and 1.29), can be determined by measuring the osmotic pressure π.

Fig. 2.2 also indicates the possibility of a diffusible component (3) being present. Since (2) is non-diffusible it is, in the experiments of major interest to us here, the biological macromolecule under investigation and (3) may well be a simple salt or buffer. From the theoretical point of view, as Scatchard has pointed out (1946), it is worth noting that interpretation is much simpler if (3) is a simple salt, such as sodium chloride, the pH being adjusted by addition of acid or alkali, rather than a buffer containing mixtures of ions of more than unit charge in ratios not always exactly determinable.

At constant temperature, even without the application of any extra pressure to the liquid on either side of the membrane (Fig. 2.2), it is possible for the chemical potential of (3) to adjust itself by diffusion so that $\mu_3^{in} = \mu_3^{out}$, since it can thereby alter the relative concentration of its constituent ions on each side of the membrane over a wide range, subject only to the laws of electroneutrality. As we have seen such a range of adjustment of its mole fraction by diffusion is not open to the solvent (1) which can only come to equilibrium by applying the pressure π. This extra pressure which allows (1) to come to equilibrium also alters the chemical potential of (3) somewhat, since $(\partial \mu_3/\partial P)_T = \bar{V}_3$. However, $\bar{V}_3$ is usually so small (~ 0.1 litres; see Tanford 1961, p. 225) that its effect on the activity is less than 0.05% and concentrations of (3) can usually be ignored, although an appropriate term must be included in any exact theoretical equations. In a dialysis experiment, when the usual aim is to allow redistribution of small diffusible solutes, like (3) in Fig. 2.2, equilibrium is only attained if the dialysis bag, through its distension, can exert the osmotic pressure on the macromolecular solution inside it.

The special virtues of determining osmotic pressure as a means of obtaining $(\mu_1 - \mu_1^0)$ and so M_2, can only be fully apparent when further equations have been developed in the next section. These show how much more experimentally sensitive than the other colligative methods (freezing point and vapour pressure lowering, and elevation of boiling point) is the osmotic pressure to the presence of a given number of solute molecules and so how much more useful it is for macromolecular solutions which are always relatively dilute in terms of *number* of molecules per unit volume. Moreover, the other colligative methods are also sensitive to the number per unit volume of all solute species, including the ions of buffers and salts. Since these are usually more numerous, the effect of a macromolecular solute can only be relatively very small and these other methods are of no use for examining aqueous solutions of biological macromolecules. However, the presence in osmotic studies of a membrane permeable to the small solute species allows the contribution of the macromolecular solute (2) to the decrement $(\mu_1 - \mu_1^0)$ to be effectively determined, provided the experiments are designed in the appropriate way (see Section 1.2). For these two reasons the measurement of osmotic pressure is the only one of the colligative methods used in practice for studying biological macromolecules in solutions in water, where other counter-ions, at least, are inevitably present; as we have seen earlier (p. 12

2.3. Neutral macromolecular solutes in two-component systems

(a) *Elementary theory and the van't Hoff equation*

If the solute (2) is an uncharged macromolecule and the only other component is solvent (1), then equations (2.3) and (1.29) can be combined thus:

$$-\pi \bar{V}_1 = \mu_1 - \mu_1^0 = -RT\bar{V}_1^0 c_2\left[\frac{1}{M_2} + Bc_2 + Cc_2^2 + \ldots\right],$$

so

$$\frac{\pi}{c_2} = RT\frac{\bar{V}_1^0}{\bar{V}_1}\left[\frac{1}{M_2} + Bc_2 + Cc_2^2 + \ldots\right].$$

$\bar{V}_1^0$ is the volume per mole of pure solvent and $\bar{V}_1$ is the partial molar volume of the solvent in the macromolecular solution. If the solution is dilute $\bar{V}_1 = \bar{V}_1^0$ so

$$\frac{\pi}{c_2} = RT\left[\frac{1}{M_2} + Bc_2^2 + Cc_2^2 + \ldots\right], \quad (2.4a)$$

or

$$\frac{\pi}{c_2} = \frac{RT}{M_2}[1 + B'c_2 + C'c_2^2 + \ldots], \quad (2.4b)$$

or

$$\frac{\pi}{c_2} = RT\left[\frac{1}{M_2} + \frac{V_m^0}{2M_2^2}\beta_{22}c_2\right]. \quad (2.4c)$$

The parallel relation which results from the use of the osmotic coefficient expression (1.31) of non-ideality is

$$\mu_1 - \mu_1^0 = -\pi\bar{V}_1 = \phi RT \ln x_1, \quad (2.5)$$

whence

$$\phi = \pi/\pi_{\text{ideal}}, \quad (2.6)$$

where

$$\pi_{\text{ideal}} = -\frac{RT}{\bar{V}_1}\ln x_1,$$

and is the osmotic pressure which would be exerted by an ideal solution of the same composition. Eqn (2.6) now explains the name 'osmotic coefficient' and at the same time provides a *definition* of an osmotic coefficient, ϕ_c, on the concentration scale in which

$$\pi = \phi_c . RTc_2/M_2 \quad (2.7)$$

and

$$\pi_{\text{ideal}} = RTc_2/M_2, \quad (2.8)$$

which is the equation derived by van't Hoff and is, of course, simply eqn (2.4a) written for ideal solutions when B and C are zero.

THEORY OF OSMOTIC PRESSURE

The van't Hoff equation (2.8) is also the limiting form as $c_2 \to 0$ of eqn (2.4a) which applies to *non-ideal* solution. This latter equation is usually plotted as π/c_2 against c_2, when the limiting intercept is RT/M_2 (but see Section 3.10). Such a plot is shown schematically in Fig. 2.3 as a straight line for the common case when the third virial coefficient $C = 0$ and the second virial coefficient $B \neq 0$. The slope then gives a value for B provided there is no association or dissociation reaction occurring in the system, so that the effective molecular weight does not change with concentration.

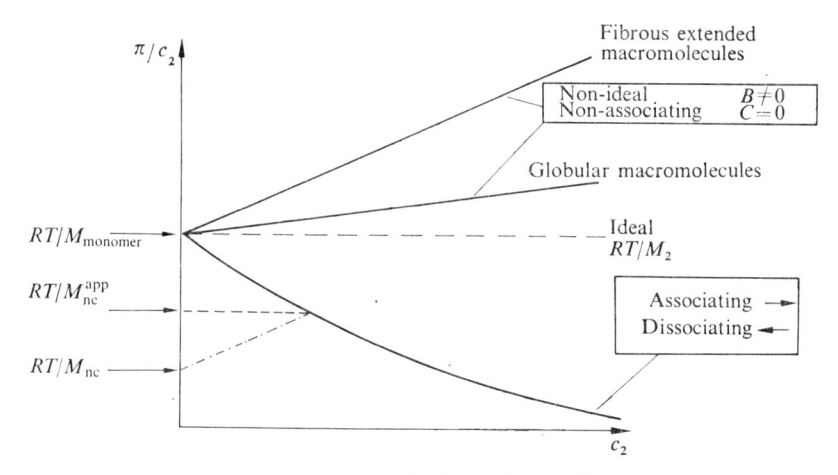

FIG. 2.3. Types of π/c_2 against c_2 plots.

Other methods of extrapolating results to infinite dilution, so as to obtain the van't Hoff relation even though the solution is non-ideal, are often empirically useful, but the form in eqn (2.4a) and Fig. 2.3 is the one most easily related to the general theory of solutions already outlined in Chapter 1.

Instead of using the form of eqn (1.29) for $(\mu_1 - \mu_1^0)$, this difference could, for ideal dilute solutions of two-components simply be written

$$\mu_1 - \mu_1^0 = RT \ln x_1 = RT \ln(1-x_2) \simeq -RTx_2;$$

hence, by eqn (2.3)

$$\pi \bar{V}_1 = RTx_2 = RTn_2/(n_1+n_2), \qquad (2.9)$$

where n_1 and n_2 are the number of moles of (1) and (2) present in the solution. Since $n_1 \gg n_2$ and $n\bar{V}_1 = V$, the actual volume of the solution, this reduces to

$$\pi V = n_2 RT, \qquad (2.10)$$

in which form it appears to be analogous to the perfect gas law ($PV = nRT$). This relationship was discovered by the pioneers of osmotic pressure studies (for a review see Glasstone 1940, ch. IX) who found that the proportionality constant between π and the product $(n_2/V)T$ was in fact the gas constant R. This striking fact has sometimes led to postulates of 'mechanisms' of osmotic pressure in which solute molecules are conceived as exerting an actual 'osmotic' pressure on the semi-permeable membrane equal to that exerted by the same concentration of molecules in a perfect gas on its containing walls. The relationship however arises directly from the parallels in the thermodynamic relationships and should not be interpreted in this molecular mechanistic sense, since the osmotic pressure is in fact a property ensuring equilibrium of the *solvent* and solute, and has its effect only via its reduction of the chemical potential of this solvent.

According to eqn (2.10) 1 mole of component (2) dissolved in 22.4 l. of solvent would exert an osmotic pressure π of 1 atmos (760 mm Hg) at 0°C, if the solution is ideal. Thus

$$R = 76 (\text{cm}) \times 13 \cdot 6 \,(\text{grams. Hg/ml}) \times 22 \cdot 4 \,(\text{litres/mole})/273$$

$$= 84 \cdot 8 \,\text{l} \,(\text{cm water}) \,\text{mol}^{-1} \,\text{deg}^{-1},$$

and, more usefully, a one molar solution of (2), i.e. 1 mole per litre of solution, has an osmotic pressure at 25° of $\pi = 84 \cdot 8 \times 295 = 2 \cdot 5 \times 10^4$ cm water. A typical biological macromolecule would have a molecular weight in the range 10^4 to 10^5 and solutions of concentration about 1 per cent, i.e. 10 grams per litre of solution would be convenient to handle. Such solutions would therefore have molarities of 10^{-3} to 10^{-4} and thus osmotic pressures at 25°C, if ideal, in the range $(2 \cdot 5 \times 10^4) \times (10^{-3} \text{ to } 10^{-4})$, i.e. 25 to 2·5 cm of water, which are readily determinable. This may be contrasted with the other colligative properties. For example the relative lowering of vapour pressure $(\Delta p/p°)$ of water in similar aqueous solutions would be equal to the mole fraction of solutes, namely $1 \times 10^{-3}/(55 \cdot 5 + 1 \times 10^{-3})$ to

$$1 \times 10^{-4}/(55 \cdot 5 + 1 \times 10^{-4}),$$

which is $1 \cdot 8 \times 10^{-3}$ per cent to $1 \cdot 8 \times 10^{-4}$ per cent and would be too small to be detectable. Similarly, the freezing point depression of water is approximately 1·9° for each mole of solute dissolved per litre, so that a 1 per cent solution of macromolecules of molecular weights in the range 10^4 to 10^5 would lower the freezing point by $1 \cdot 9 \times 10^{-3}$ to $1 \cdot 9 \times 10^{-4}$°C, which is again too small to be measurable with any accuracy.

By the application of special techniques for measuring very small differences of vapour pressure or very small temperature differences the corresponding colligative methods have been made applicable to the study of synthetic macromolecules in non-aqueous solvents (Weissberger, 1961, VI. I, IX), when the absence of small contaminating molecules and ions can be ensured. But the much greater sensitivity of the measurement of the osmotic pressure, just demonstrated, ensures its superiority among the colligative methods applied to macromolecules and is the only one of these methods applicable to charged biological macromolecules in aqueous solutions, when small ions can never be completely eliminated.

The foregoing treatment has not been entirely rigorous and a more exact thermodynamic treatment is now necessary in order to be able to deal properly with charged macromolecules and multicomponent systems and this follows in the next section.

(b) *Exact treatment*

Following the notation of Chapter 1, the Gibbs–Duhem condition for equilibrium in a mixture of components J in the absence of any gravitational or electrical fields, at constant temperature is

$$V \, dP = \sum_J n_J \cdot d\mu_J. \tag{1.11}$$

If only two components are present (1 = diffusible solvent; 2 = non-diffusible solute) and the extensive quantities (V, n_J) are referred to an amount of solution containing 1000 g of solvent (1), this equation becomes

$$V_m \cdot dP = \sum_J m_J \cdot d\mu_J = m_1 \, d\mu_1 + m_2 \, d\mu_2 \tag{2.11}$$

The condition of osmotic equilibrium is that the chemical potential, μ_1, of the diffusible component, solvent (1), is maintained constant (at its value in the pure solvent in the 'outside' solution (Fig. 2.2) under a constant pressure); so

$$d\mu_1 = 0$$

and

$$V_m \cdot dP = m_2 \, d\mu_2 \quad \boxed{\text{Constant } \mu_1, T} \tag{2.12a}$$

The constancy of T and μ_1 under which the differentials in eqn (2.12) are performed can be expressed by writing it as

$$\left(\frac{\partial \mu_2}{\partial P}\right)_{\mu_1, T} = \frac{V_m}{m_2}. \tag{2.12b}$$

The experimental variables of this system under the conditions of an osmotic equilibrium experiment are the pressure (P) applied to the solution and the concentration (molality, m_2) of component (2), which is, for our present purposes, a macromolecule, though the following treatment is quite general. (Note that the molality, m_2, is enough to specify composition exactly.) We require the relation between these experimental variables, for example, an expression for $(\partial P/\partial m_2)_{\mu_1, T}$. This may be obtained as follows. Since P and m_2 are the variables of the system (at constant μ_1 and T), it is always true that

$$d\mu_2 = \left(\frac{\partial \mu_2}{\partial m_2}\right)_{T, P} \cdot dm_2 + \left(\frac{\partial \mu_2}{\partial P}\right)_{m_2, T} \cdot dP$$

$$= RT a_{22} \cdot dm_2 + \bar{V}_2 \cdot dP,$$

following the definitions of Chapter 1 (eqn 1.12, 1.33). Differentiation of this equation with respect to m_2, under conditions of constant μ_1 and T, yields

$$\left(\frac{\partial \mu_2}{\partial m_2}\right)_{\mu_1, T} = RT a_{22} + \bar{V}_2 \left(\frac{\partial P}{\partial m_2}\right)_{\mu_1, T}. \tag{2.13}$$

The l.h.s. is equal to

$$\left(\frac{\partial \mu_2}{\partial P}\right)_{\mu_1, T} \cdot \left(\frac{\partial P}{\partial m_2}\right)_{\mu_1, T}.$$

Substitution of the equilibrium value of the first of these partial differentials from the condition (2.12b) and multiplication by m_2 transforms eqn 2.13 to

$$V_m \cdot \left(\frac{\partial P}{\partial m_2}\right)_{\mu_1, T} = RT m_2 a_{22} + m_2 \bar{V}_2 \left(\frac{\partial P}{\partial m_2}\right)_{\mu_1, T}$$

$$\therefore \quad \boxed{\left(\frac{\partial P}{\partial m_2}\right)_{\mu_1, T} = \frac{RT m_2 a_{22}}{(V_m - m_2 \bar{V}_2)}} \quad \begin{array}{l}\text{Equilibrium} \\ \text{Constant } \mu_1 \\ \text{Constant } T\end{array} \tag{2.14}$$

Eqn (2.14) expresses exactly how the pressure (P) applied to the solution at constant temperature must change with the molality of the solute (2) in order to maintain constant the chemical potential (μ_1) of the solvent (1), i.e. osmotic equilibrium. Thus the pressure P in this expression is the *osmotic pressure* π and the partial differential can be replaced by $d\pi/dm_2$, since the conditions of constancy of μ_1 are automatically implied in this replacement. Other substitutions in the R.H.S. of eqn (2.14) can be made, some accurate, others less so.

Thus, it is exactly true that

$$V_m - m_2 \bar{V}_2 = m_1 \bar{V}_1 = \frac{1000}{M_1} \cdot \bar{V}_1,$$

where $\bar{V}_1$, is the partial molal volume of the solvent in the solution under pressure π and m_1 is now the molality of (1) in a solution containing 1000 g of it. Hence eqn (2.14) becomes

$$\boxed{\frac{d\pi}{dm_2} = \frac{RT M_1 m_2 a_{22}}{1000 \bar{V}_1}} \quad \begin{array}{l}\text{Equilibrium} \\ \text{Constant } \mu_1, T\end{array} \tag{2.15}$$

This is an exact expression for the change of osmotic pressure with solute concentration for a two-component system. Since the solutions are dilute enough and the compressibility terms (e.g. $1/\bar{V}_1(\partial \bar{V}_1/\partial P)_T$) are usually negligibly small (see Casassa and Eisenberg 1964, p. 297 for the detailed requirements) the denominator $(V_m - m_2 \bar{V}_2)$ in eqn (2.14) may be replaced by V_m^0, the limiting volume (in ml) of the solution containing 1000 g of solvent

(1) as $m_2 \to 0$, and the exact equation (2.15) assumes the only slightly less exact form

$$\frac{d\pi}{dm_2} = \frac{RT}{V_m^0} \cdot m_2 a_{22} \qquad \begin{array}{l}\text{Equilibrium}\\ \text{Constant } \mu_1, T \\ \text{Dilute} \\ \text{Compressibility ignored}\end{array} \qquad (2.16)$$

The expressions in eqns (1.34, 1.35) of Chapter 1 for a_{22} when the solutions are *not ideal* and (2) is uncharged may now be substituted to give

$$\frac{d\pi}{dm_2} = \frac{RT}{V_m^0}(1+\beta_{22}m_2) \qquad (2.17a)$$

and, more usefully,

$$\frac{d\pi}{dm_2} = \frac{RT}{V_m^0}(1+2BM_2c_2+3CM_2c_2^2+\ldots). \qquad (2.17b)$$

Substitution of $m_2 = c_2 V_m^0/M_2$ (with the same degree of exactitude as the substitution of V_m^0 in eqn (2.16) for the denominator in eqn (2.14)), integration with respect to c_2 (from 0 to c_2 and π from 0 to π), and division by c_2 gives again

$$\frac{\pi}{c_2} = RT\left[\frac{1}{M_2}+Bc_2+Cc_2^2+\ldots\right] \qquad \begin{array}{l}\text{Equilibrium}\\ \text{Constant } \mu_1, T \\ \text{Dilute} \\ \text{Compressibility ignored}\end{array} \qquad (2.4a)$$

though its derivation and relationships to other forms have now been given a more general basis. In practice, even for a non-ideal solution, at the concentrations of biological macromolecules usually convenient for study, the third virial coefficient C, in eqn (2.4a) is negligible. However, this coefficient plays a significant part in concentrated solutions of charged macromolecules under, for example, gel conditions (Nichol, Ogston, and Preston 1967). As before, if the solution is ideal the virial coefficients B and C are negligibly small and this equation reduces to the van't Hoff equation (2.8), which we see from (2.17a), when $\beta_{22} = 0$, can also be written with molalities as

$$\boxed{\frac{V_m^0}{RT} \cdot \frac{d\pi}{dm_2} = 1} \quad \text{or} \quad \boxed{\pi = \frac{V_m^0}{RT} \cdot m_2} \qquad \begin{array}{l}\text{Equilibrium}\\ \text{Ideal (van't Hoff)} \\ \text{Constant } \mu_1, T \\ \text{Dilute} \\ \text{Compressibility ignored}\end{array} \qquad (2.18)$$

On a weight scale, which has the advantage that composition is independent of temperature and in which $w_2 = m_2 M_2$ is the weight of (2) in 1000 g of (1), the van't Hoff equation assumes the form

$$\frac{\pi}{w_2} = \frac{RT}{V_m^0} \cdot \frac{1}{M_2} \qquad \begin{array}{l}\text{Equilibrium}\\ \text{Ideal (van't Hoff)}\\ \text{Constant } \mu_1, T\\ \text{Dilute}\\ \text{Compressibility ignored}\end{array} \qquad (2.19)$$

2.4. Charged macromolecular solutes in two-component systems

The form of the osmotic pressure relationships in the previous section which is the most useful in practice, namely the virial equation (2.4a) has been derived from eqn (2.16) on the assumption, in substituting for a_{22}, that the solute component (2) is uncharged. Most biological macromolecules are charged and are inevitably accompanied by their corresponding neutralizing counter-ions, so it must be represented as PX_{z_2} where z_2 is the algebraically positive charge on the macro-ion P and X is a univalent counter ion (anion if z_2 is positive). The expressions (1.34a, 2.35a) containing the extra factor (z_2+1) must then be used for a_{22} and in eqn 2.16. In particular eqn 2.4a has to be replaced by

$$\frac{\pi}{c_2} = RT(z_2+1)\left[\frac{1}{M_2} + Bc_2 + Cc_2 + \ldots\right] \qquad (2.20)$$

Thus extrapolation to $c_2 = 0$ does not remove the common factor (z_2+1) which means that the osmotic pressure measured with only water outside the membrane can no longer give M_2 unless z_2 is accurately known. In practice, this means that molecular weights of charged macromolecules cannot be measured in such two-component systems, which is directly obvious since the colligative method, such as an osmotic pressure measurement, counts all the (z_2+1) ionic species per mole of component (2) present in the solution and will give the average molecular weight (see Section 2.6 below) of one macro-ion and small ions together. If component (2) could be defined so that it was uncharged then it might be thought this problem could be avoided. This is achieved by the definition of Scatchard (Section 1.4) but involves the use of a third component of neutral salt. As we have seen, the Scatchard definition of component (2) can be generalized to the more fundamental form of Casassa and Eisenberg, for which $a_{23} = 0$, but the important point to note is that both require the presence of a third salt component. Once such a component is present the Donnan equilibrium of this salt will establish itself and the imbalance of salt concentration thereby produced across the membrane will itself contribute to the osmotic pressure. The appropriate relationships for such systems must now be derived on the basis of an exact thermodynamic treatment in order to determine if osmotic pressure measurements can still yield the molecular weights of charged macromolecules as it does of neutral ones.

2.5. Three-component systems

The three components of the solution are two diffusible components, solvent (1) and neutral salt (3); and one non-diffusible component (2), the macromolecule in the cases of interest here. This solution is taken to be in osmotic equilibrium across a membrane with a solution containing only (1) and (3). The starting point is again the Gibbs–Duhem condition for equilibrium at constant temperature.

$$V_m . dP = \sum_J n_J . d\mu_J, \quad \boxed{\text{Constant } T} \quad (1.11)$$

which becomes, when referred to an amount of solution containing 1000 g of solvent (1),

$$V_m . dP = \sum_J m_J . d\mu_J$$
$$= m_1 \, d\mu_1 + m_2 \, d\mu_2 + m_3 \, d\mu_3.$$

The condition of osmotic equilibrium is that the chemical potentials of both components (1) and (3) remain constant and equal to their values in the mixture of (1) and (3) in the 'outside' solution (Fig. 2.2) under a constant pressure; so

$$d\mu_1 = 0 = d\mu_3$$

and

$$V_m . dP = m_2 \, d\mu_2 \quad \boxed{\text{Constant } \mu_1, \mu_3, T}$$

The constancy of μ_1, μ_3 and T under which the differentials occur in this equation can be expressed by writing it as

$$\left(\frac{\partial \mu_2}{\partial P}\right)_{\mu, T} = \frac{V_m}{m_2}, \quad (2.21)$$

where subscript μ denotes constancy of chemical potentials of all components other than (2). The experimental variables of interest at osmotic equilibrium are again the pressure (P) applied to the solution and the concentration (molality m_2) of component (2) and, as before, we require $(\partial P/\partial m_2)_{\mu, T}$. For *any* change in the chemical potential of (2) at constant temperature

$$d\mu_2 = \left(\frac{\partial \mu_2}{\partial P}\right)_{T, m_2, m_3} . dP + \left(\frac{\partial \mu_2}{\partial m_2}\right)_{T, P, m_3} . dm_2 + \left(\frac{\partial \mu_3}{\partial m_3}\right)_{T, P, m_2} . dm_3$$
$$= \bar{V}_2 . dP + RT a_{22} . dm_2 + RT a_{23} . dm_3,$$

following the definition of Chapter 1 (eqns 1.12, 33). Note that the molalities m_2 and m_3 are together sufficient to specify the composition also with respect to component (1), for they both correspond to 1000 g of (1).

This equation can be written as

$$-(d\mu_2 - \bar{V}_2 \, dP)/RT + a_{22} \, dm_2 + a_{23} \, dm_3 = 0,$$

and a similar equation can be written for $d\mu_3$:

$$-(d\mu_3 - \bar{V}_3 \, dP)/RT + a_{32} \, dm_2 + a_{33} \, dm_3 = 0.$$

Multiplication of the first of these two by a_{33}, of the second by a_{23}, and subtraction of the two resulting equations eliminates dm_3 and gives

$$RT \cdot dm_2(a_{22}a_{33} - a_{23}^2) = (d\mu_2 - \bar{V}_2 \, dP)a_{33} - (d\mu_3 - \bar{V}_3 \, dP)a_{32},$$

since $a_{23} \equiv a_{32}$. Differentiation with respect to m_2, under conditions of constant μ_1, μ_3 and T (the condition of osmotic equilibrium) gives

$$\left(\frac{\partial \mu_2}{\partial m_2}\right)_{\mu,T} - \left(\frac{\partial P}{\partial m_2}\right)_{\mu,T}\left(\bar{V}_2 - \bar{V}_3 \frac{a_{32}}{a_{33}}\right) = RT\left(a_{22} - \frac{a_{23}^2}{a_{33}}\right)$$

since $d\mu_3 = 0$. But the first term on the L.H.S. is equal to

$$\left(\frac{\partial \mu_2}{\partial P}\right)_{\mu,T}\left(\frac{\partial P}{\partial m_2}\right)_{\mu,T}$$

and substitution into this general expression of the value of the first of these partial differentials from the equilibrium condition (2.21) and multiplication by m_2 yields

$$\left(\frac{\partial P}{\partial m_2}\right)_{\mu,T}\left[V_m - m_2\left(\bar{V}_2 - \bar{V}_3 \frac{a_{32}}{a_{33}}\right)\right] = RTm_2\left(a_{22} - \frac{a_{23}^2}{a_{33}}\right) \quad (2.22)$$

This is the exact expression for the relation between m_2 and the pressure P, which is now the osmotic pressure π, for it is that pressure which will maintain μ_1 and μ_3 constant at constant T. So, as before, the partial differential in this equation can be written as $d\pi/dm_2$. As mentioned in Section 2.2, $\bar{V}_3$ is very small and on the same basis as for eqn (2.16) the quantity in the square bracket can be replaced for dilute solutions by V_m^0 (Casassa and Eisenberg 1964, pp. 297 ff.). So the completely exact eqn (2.21) becomes for three-component systems:

$$\boxed{\frac{d\pi}{dm_2} = \frac{RT}{V_m^0} \cdot m_2\left(a_{22} - \frac{a_{23}^2}{a_{33}}\right),}$$

Equilibrium
Constant μ_1, μ_3 and T
Dilute
Compressibility
and $\bar{V}_3$ ignored.

(2.23a)

which is the analogue of (2.16) for two-component systems. Since

$$m_2 = c_2 V_m^0/M_2,$$

by the same approximations as before (see text preceding eqn (2.16)) this can be written

$$\boxed{\frac{d\pi}{dc_2} = \frac{RT}{M_2} m_2 \left(a_{22} - \frac{a_{23}^2}{a_{33}} \right).}$$

Equilibrium Constant μ_1, μ_3, T
Dilute
Compressibility and $\bar{V}_3$ ignored. (2.23b)

The last part of this expression has already been evaluated for two conditions.

(a) A three-component system containing only two kinds of small ions at ideal Donnan equilibrium (activity coefficients of small ions equal to unity) with the Scatchard definition of component (2), i.e. ν of the Casassa and Eisenberg definition is $-z_2/2$ and $\epsilon \to 1$, as in Section 1.5.

Substitution of the expression giving $m_2(a_{22}-a_{23}^2/a_{33})$ for ideal Donnan equilibrium conditions from eqn (1.50b) into the general expression (2.23b) for $d\pi/dc_2$ and integration with respect to c_2 (from 0 to c_2 and π from 0 to π) yields

$$\frac{\pi}{c_2} = RT\left[\frac{1}{M_2} + \frac{V_m^0}{2M_2^2}\left(\underline{\frac{z_2^2}{2m_3}} + \beta_{22} - \frac{\beta_{23}^2 m_3}{2+\beta_{33}m_3}\right)c_2\right], \quad \begin{array}{|c|} \hline \text{Donnan} \\ \nu = -z_2/2 \\ \epsilon \to 1 \\ \hline \end{array} \quad (2.24a)$$

which may be written as

$$\frac{\pi}{c_2} = RT\left[\frac{1}{M_2} + B'c_2\right], \quad (2.24b)$$

from eqn (1.51a). The first underlined term (eqn (2.24a)) in the second virial coefficient B' represents the contribution to the total osmotic pressure of the difference between the total molalities of small diffusible ions inside and outside the membrane, namely $(m_+^{in}+m_-^{in}-m_+^{out}-m_-^{out})$ and may be called the 'Donnan term'.

For this imbalance of ionic composition is given, for Donnan conditions ($\nu = -z_2/2$) by eqn (1.49),

$$\Gamma_{32} = \frac{\Delta m_3}{\Delta m_2} = \frac{m_3^{in}-m_3^{out}}{m_2^{in}-m_2^{out}} = +\frac{z_2^2 m_2}{4m_3},$$

where m_2 and m_3 both refer to a macromolecular solution. Hence $\Delta m_3 = z_2^2 m_2^2/4m_3$, since $m_2^{out} = 0$, and by substitution of $m_2 = V_m^0 c_2/M_2$ and application of the van't Hoff law the Donnan term just noted is obtained directly. Even when the macromolecule behaves ideally, so that $\beta_{22} = 0 = \beta_{23}$, this term is still present as a contribution to B' but disappears in the extrapolation of π/c_2 to infinite dilution. In practice it is kept as small as possible by making the concentration of neutral salt m_3 as large as possible, but in the most accurate work (e.g. Adair 1925, 1928) it is advisable to calculate this Donnan contribution to the total osmotic pressure on the above basis or to obtain the difference in ionic concentrations indirectly by means of membrane potential measurements (Adair 1961).

(b) A three-component system at osmotic equilibrium, not under ideal Donnan conditions, i.e. with an imbalance of uni–univalent electrolyte so that $v \neq -z_2/2$. The quantity $z_2^2/2m_3$ in the Donnan term of the previous paragraph is then replaced by $\sum_i v_{2i}^2/m_i$, where v_{2i} is defined, as in Section 1.4, as the moles of diffusible ionic species i included in the macromolecular component (2) and the contribution of the ionic imbalance to the osmotic pressure is then

$$\pi_{\text{Donnan}}^{\text{ionic}} = \frac{RT m_2^2}{2 V_m^0} \sum_i \frac{v_{2i}^2}{m_i} = \frac{RT V_m^0 c_2^2}{2 M_2^2} \sum_i \frac{v_{2i}^2}{m_i} \quad (2.25)$$

By substituting eqn (1.51b) for $m_2(a_{22}-a_{23}^2/a_{33})$ into the osmotic pressure expressions (e.g. eqn 2.23b) and integrating as before, eqn (2.24a) becomes

$$\frac{\pi}{c_2} = RT\left[\frac{1}{M_2} + \frac{V_m^0}{2M_2^2}\left(\sum_i \frac{v_{2i}^2}{m_i} + \beta_{22} - \frac{a_{23}^2}{a_{33}}\right)c_2\right] \quad (2.26a)$$

or, by eqns (1.50d) and (1.51b),

$$\frac{\pi}{c_2} = RT\left[\frac{1}{M_2} + B''c_2\right]. \quad (2.26b)$$

If the Casassa–Eisenberg definition of component (2) is employed (the starred quantities of Section 1.5 for which $a_{23}^* = 0$) then the last interaction term of eqn (2.26a) disappears giving

$$\frac{\pi}{c_2^*} = RT\left[\frac{1}{M_2^*} + \frac{V_m^0}{2M_2^{*2}}\left(\sum_i \frac{v_{2i}^{*2}}{m_i} + \beta_{22}^*\right)c_2^*\right], \quad (2.27a)$$

or

$$\frac{\pi}{c_2^*} = RT\left[\frac{1}{M_2^*} + B^* c_2^*\right], \quad (2.27b)$$

by eqn (1.54).

This formulation of macromolecular component (2*) thus simplifies the final expressions by eliminating all the terms representing interaction between the macromolecule and small diffusible ions (a_{23}, β_{23}) and also the non-ideality term for the salt (a_{33}). But this, and indeed all definitions of macromolecular component (2), must still leave in the final expression for the total osmotic pressure the contribution to it of the ionic imbalance. Whether or not the osmotic equilibrium is of the ideal Donnan form, as in (a) above, there is still the same need to reduce the Donnan term, eqn (2.25), to negligible proportions or to estimate it indirectly: the former is the procedure usually followed.

Extrapolation of the osmotic pressure results to infinite dilution, whether as π/c_2 vs. c_2 or in some other form, is an extrapolation both for two-component and three-component solutions to conditions where the van't Hoff relation applies in the form which involves either concentrations (eqn 2.8) or molalities (eqn 2.18). These two forms of the van't Hoff relation state that the osmotic pressure is directly proportional to the quantities c_2/M_2 or to $m_2 = 1000\, w_2/M_2$, respectively; these are ratios of analysed *weights* of component (2), per unit volume of solution or per unit weight of solvent, to the molecular weight of (2). As the account in the last paragraphs of Chapter 1 and eqns (1.59, 1.60) show in more detail, this means that the entity whose molecular weight is determined in the osmotic pressure experiment is that with respect to which the analytical determinations of weights of (2) were calibrated—even for three- and multi-component solutions with their inherent complications concerning the definition of (2).

2.6. Mixtures of macromolecules

All the colligative methods have the virtue that if applied to mixtures of macromolecules they can, in principle yield the number average molecular weight, M_n. As has been already mentioned, of these only the osmotic pressure method is, in practice, applied to the study of solutions of biological macromolecules, although the measurement of vapour pressure lowering and of freezing point depression can be applied (Weissberger 1959, Vol. I, Ch. VII, IX) to the study of synthetic macromolecules in non-aqueous solvents. That the osmotic pressure determines M_n, when the macromolecules are not involved in any chemical reaction and conditions are thermodynamically ideal (either actually or by suitable extrapolation), may be proved as follows. Suppose that the mixture contains a weight concentration of $c_1, c_2, \ldots c_i \ldots$ (g ml^{-1}) of macromolecules of molecular weights M_1, $M_2, \ldots M_i \ldots$, respectively, and let the contribution of each species to the total osmotic pressure (π), under the conditions just specified be π_1, $\pi_2, \ldots \pi_i \ldots$, respectively, then

$$\pi = \sum_i \pi_i = RT \sum_i c_i/M_i$$

But $\sum c_i = c$, the total weight concentration in g ml^{-1} and the mean molecular weight, $\bar{M}_{\mathrm{OP}}$, determined from osmotic pressure measurements, is given by

$$\pi = RT \cdot \frac{c}{\bar{M}_{\mathrm{OP}}}$$

Comparison of these two equations shows at once that

$$\bar{M}_{\mathrm{OP}} = \frac{c}{\sum_i c_i/M_i} = \frac{\sum c_i}{\sum_i c_i/M_i} = M_n, \qquad (2.28)$$

for this is the definition of a number average molecular weight, M_n, and this must be the quantity obtained by applying the van't Hoff equation or by extrapolating the virial eqn (2.4a) to $c_2 = 0$. This result is not surprising since, as was pointed out earlier, the colligative methods effectively count the number of solute molecules per unit volume of solution (or weight of solvent) and if the weight concentrations (c_2 or w_2) are known, the average molecular weight so obtained must be M_n, since this is defined so as to be the result of just such a process. Because of this, measurement of the osmotic pressure should be particularly applicable to the study of processes of degradation or dissociation of biological macromolecules since the osmotic pressure increases in proportion to the increase in the number of macromolecules present (See section 2.8).

2.7. The significance of the second virial coefficient

This coefficient, on the scale of concentration in g/ml, is denoted by B (with various subscripts or superscripts) and is derived from the slope of the plot of π/c_2 against c_2, after division by RT.† For neutral macromolecules the equation of this plot is

$$\frac{\pi}{c_2} = RT\left[\frac{1}{M_2} + \frac{V_m^0}{2M_2^2}\beta_{22} \cdot c_2\right] \qquad (2.4c)$$

so that then

$$B = V_m^0 \beta_{22}/2M_2^2;$$

and for charged macromolecules in the presence of a third salt component (3), the corresponding equation is

$$\frac{\pi}{c_2} = RT\left[\frac{1}{M_2} + B''c_2\right] \qquad (2.26b)$$

in which the second virial coefficient is

$$B'' = \frac{V_m^0}{2M_2^2}\left[\underbrace{\sum_i \frac{v_{2i}^2}{m_i}}_{\text{I}} + \underbrace{\beta_{22}}_{\text{II}} - \underbrace{\frac{a_{23}^2}{a_{33}}}_{\text{III}}\right]. \qquad (1.51b)$$

The meanings of the three contributions, which are inside the square bracket and are denoted by I, II, and III, are worth examining further.

I. The contribution I to the second virial coefficient is the so-called Donnan term and is expressed in eqn (1.51b) in its most general form, which simplifies to $z_2^2/2m_3$ when only two kinds of monovalent ion are present, as in eqn (2.24a). It represents the contribution, π_i, to the total osmotic pressure π of the imbalance in the concentrations of the small diffusible ions on each side of the membrane as a result of the setting up of the Donnan equilibrium of these ions. Unlike the other two contributions, II and III, it does not vanish when solutions are dilute enough for *all* activity coefficients, including those of the small ions, to be unity. It only becomes negligible when the total salt

† This division by RT will be taken for granted in the subsequent discussion.

concentration becomes relatively large (i.e. when m_i in its denominator is large) so that the concentration of diffusible salt components is increased; and/or when the squares of the quantities ν_{2i}, which represent the number of moles of the various diffusible ions i which are 'bound' to the macromolecule, is reduced to zero. If the macromolecule 'binds' only H⁺ ions, then ν_{2,H^+}^2 becomes zero at the iso-electric point, but, if other ions are 'bound' as well,

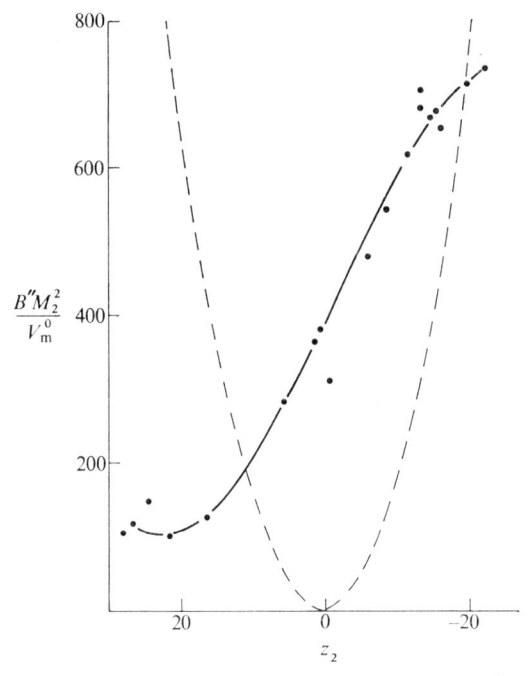

FIG. 2.4. Relation between the second virial coefficient ($B''M_2^2/V_m^0$ in our notation ("BW₂" in the notation of the original authors) and the charge (z_2) per molecule of bovine serum albumin. ● Observed points from osmotic pressure measurements on bovine serum albumin in 0·15 M NaCl; z_2 varied by adjusting pH.—drawn to fit these points. - - - - Plot of $z^2/4m_3$ for ideal Donnan contribution to the second virial coefficient (with z_2 determined only by H⁺ ion binding). (From Scatchard, Batchelder, and Brown, 1946, Fig. 9.)

the minimum in the quantity (I) may occur at a pH removed from the iso-electric point when the net charge is indeed zero but there is a finite charge on groups combining with H⁺ ions. This is illustrated in Fig. 2.4, which is taken from the comprehensive measurements of Scatchard *et al.* (1946) on the osmotic pressure of aqueous solutions of bovine serum albumin in 0·15 M-NaCl, at 0°C, and in which the ordinate represents, in our notation, the quantity $B''M_2^2/V_m^0$. The abscissa is the charge, z_2 per molecule, calculated from the H⁺ ion dissociation curve and this figure shows that the minimum in B'' is not at the zero of the charge so calculated. The dashed line, in fact, plots $z_2^2/4m_3$ and would be followed if only an ideal Donnan term contributed to the second virial coefficient. It must be concluded that in the conditions of these experiments about 20 Cl^- ions were bound to each protein molecule so that when 20 H^+ were bound (z_2 from $H^+ = +20$) the charge was neutralized by an equal number of chloride ions to produce a minimum in this plot. This interpretation was later supported by other measurements (Scatchard et al. 1950; Edsall et al. 1950). If only one diffusible salt is present, as in the experiments depicted in Fig. 2.4, the Donnan term I together with the factor $(V_m^0/2M_2^2)$ outside the square bracket of eqn (1.51b) includes the ratio z_2^2/M_2^2 which is the square of the net charge per g of macromolecule and is independent of the molecular weight M_2.

In practice, it may not always be possible to work very close to the point where this net charge is nearly zero, because of the low solubility of most macromolecules under these conditions, and the Donnan term is kept as low as possible by having present an adequate concentration of diffusible salt, usually of about 0·1 ionic strength. Since the amount of salt needed to suppress this term depends on the actual charge it is usually wise in any practical case to calculate the contribution, π_i, of the Donnan term to the total osmotic pressure, by using the charge deduced from a combination of hydrogen ion titration curves and other ion binding parameters, if known, and the general expression

$$\pi_i = \frac{RT}{V_m^0} \cdot \frac{(m_2 z_2)^2}{2 \sum_i m_i z_i^2}, \qquad (2.2a)$$

where the m are molalities and $\sum_i m_i z_i^2$ is twice the ionic strength calculated from molalities (rather than the usual molarities). If the conditions are not designed to reduce the Donnan contribution to negligible proportions it can completely dominate the virial coefficient, to the point of rendering the contribution of the macromolecule itself too small to allow M_2 to be determined (which in the limit is the situation described in 2.4, above (eqn 2.20).

II. Term II in the sub-division of the second virial coefficient in eqn (1.51b) is, as eqn (2.4a) shows, the only one to appear in the second virial coefficient of uncharged macromolecules. It represents the variation of the activity coefficient of the macromolecule with its own concentration, that is the effect of this concentration on deviations of its free energy from the ideal value. With charged macromolecules, it does not disappear at high ratios of salt to macromolecular concentration or near to the iso-electric point. This term (II) has been much discussed in connection with the behaviour in solution of uncharged macromolecules, for it is related to their effective size and shape. The same considerations apply to solutions of charged macromolecules when Donnan effects (I) are suppressed, since (III) is either zero (through the use of 'starred' component definitions, see Section 1.6) or is negligibly small (from the presence of excess salt needed, in any case, to suppress the effect

(I)—see discussion of (III), below). The present treatment of the contribution of (II) will therefore be conducted initially in terms of solutions of neutral macromolecules in a single solvent, a two-component system, by using the following equations:

and

$$\frac{\pi}{c_2} = RT\left[\frac{1}{M_2} + Bc_2 + Cc_2^2 + \ldots\right] \quad (2.4a)$$

$$\frac{\pi}{c_2} = RT\left[\frac{1}{M_2} + \frac{V_m^0}{2M_2^2}\beta_{22} \cdot c_2\right] \quad (2.4c)$$

In Chapter 1, we saw how the deviations from the ideal values for the partial molal free energy of dilution $(\mu_1 - \mu_1^0)$ of the solvent (1) in such a two-component mixture could be represented by the second virial coefficient B, in eqn (1.29), and how the relationship of this to activity coefficients (e.g. eqns 1.30a, 1.35) would be formulated—a relationship which allows eqns (2.4a, b, and c) to be written as equivalent. Deviations from ideality were found to be much illuminated by the realization that the ideal value of the partial molal free energy of dilution of the solvent could be written thus (cf. eqn 1.23):

$$\Delta \bar{G}_1^{\text{ideal}} = \mu_1 - \mu_1^0 = \Delta \bar{H}_1^{\text{ideal}} - T \cdot \overline{\Delta S}_1^{\text{ideal}}$$

$$= 0 + RT \ln x$$

$$\simeq 0 - RTx_2, \quad \text{if } x_2 \ll 1$$

$$= 0 - RT \cdot V_1^0 \frac{c_2}{M_2}, \quad \text{by eqn (1.4)}.$$

Recalling (eqn (2.3)) that $\mu_1 - \mu_1^0 = -\pi \bar{V}_1$, it was clear that the ideal van't Hoff law for the osmotic pressure, that is zero values of the second virial coefficient, can be attributed to ideal values of the entropy of mixing (eqn 1.20) of the two components and to zero heats of mixing. However, as was pointed out in Section 1.3, macromolecules inevitably have non-ideal entropies of mixing, quite apart from any extra configurational entropy which might ensue from flexibility. This contribution to nonideality, through the $\overline{\Delta S}_1$ term, is usually called an effect of 'excluded volume',† partly by analogy with the role of the b term in the van der Waals' expression‡ for imperfect gases which also appears as a positive contribution to the second virial coefficient in an expansion‡ of PV as powers of V^{-1}, which, for a 1 mole of gas, is the same as powers of concentration. The same expression‡ also

† Often even when entropic effects of flexibility are included.
‡ The van der Waals' equation for 1 mole of gas is

$$\left(P + \frac{a}{V^2}\right)(V - b) = RT,$$

shows that the attractive forces between molecules make a negative contribution to the second virial coefficient which can exceed numerically the excluded volume effect. This is not uncommon with gases but rare with macromolecular situations since such strong attractive forces would usually cause precipitation. To describe the positive contribution to term II in the second virial coefficient as an 'excluded volume' effect and as a modification of the $\overline{\Delta S_1}$ term are complementary ways of speaking. For to obtain $\overline{\Delta S_1}$ it is necessary to develop an expression for ΔS_{mix} which is then differentiated with respect to the number (n_1) of moles of solvent (1) in the solution to give $\overline{\Delta S_1}$ (eqn 1.22). The quantity ΔS_{mix} is the entropy of n_1 moles of (1) and n_2 moles of (2) in the mixture minus their entropies in the separate state before mixing and each of these entropies is related by the Boltzmann equation ($S = (R/N_0)\ln W$) to the number of ways W of arranging the respective systems. If the solution is sufficiently dilute that one macromolecule does not affect the internal motion of another, the only factor affecting these differences in W and in the entropies is the reduction, u_2, which the presence of one macromolecule makes on the average† volume available to the next one to be added to the mixture. This reduction, u_2, is the 'excluded volume'. On this basis it can be shown (cf. Tanford 1961, Section 12a)

$$\overline{\Delta S_1} = S_1 - S_1^0 = RV_1^0 c_2 \left[\frac{1}{M_2} + \frac{N_0 u_2}{M_2^2} \cdot c_2 \right] \qquad (2.30)$$

If the segments of a neutral macromolecule are presumed to have an 'ideal' interaction with the solvent molecules, so that $\overline{\Delta H_1} = 0$, since the partial molar free energy of dilution is

$$\mu_1 - \mu_1^0 = \Delta \bar{H}_1 - T\,\overline{\Delta S_1},$$

where a is the repulsion factor and b is the excluded volume per mole of gas, i.e. b/N_0 is the volume around each molecule of gas from which the centre of any other molecular volume is excluded (N_0 is Avogadro's number). This equation can be written in the form

$$PV = RT\left[1 + \left(b - \frac{a}{RT}\right)\left(\frac{1}{V}\right) + \frac{ab}{RT}\left(\frac{1}{V}\right)^2\right],$$

when it may then be compared with

$$\pi \bar{V}_1 = -(\mu_1 - \mu_1^0) = RT\bar{V}_1^0 c_2\left[\frac{1}{M_2} + Bc_2 + Cc_2^2 + \ldots\right] \qquad (1.29) \text{ and } (2.3)$$

of the preceding solution theory; or the van der Waals' equation may be written in the form obtained by substituting M/c for V, the volume of 1 mole of gas, namely

$$\frac{P}{c} = RT\left[\frac{1}{M} + \left(b - \frac{a}{RT}\right)\frac{c}{M^2} + \frac{ab\,c^2}{RTM^3} + \ldots\right],$$

when it may then be compared with the equation for the osmotic pressure

$$\frac{\pi}{c_2} = RT\left[\frac{1}{M_2} + Bc_2 + Cc_2^2 + \ldots\right]. \qquad (2.4a)$$

† 'On the average' because each molecule is taken to exclude the centre of another like molecule from a volume of $2u_2$.

then
$$\mu_1 - \mu_1^0 = -RTV_1^0 c_2 \left[\frac{1}{M_2} + \frac{N_0 u_2}{M_2^2} \cdot c_2 \right].$$

This shows that the second virial coefficient of eqn (1.29) would, if determined by the excluded volume alone, be

$$B_{II} = \frac{N_0 u_2}{M_2^2}, \qquad (2.31)$$

where the subscript II in B_{II} refers back to the subdivision of B of eqn (1.51b) into three terms.

This expression is quite general and useful for giving an indication of the contribution of excluded volume effects to the term II in the second virial coefficient. Thus, a rigid sphere excludes the centre of any other similar sphere from the volume enclosed in a sphere of twice its own radius, and so eight times its own volume. This exclusion is that of both spheres taken together, so the excluded volume for each is, on the average, half this, namely four times the volume of each. Thus if v_2 is the specific volume (i.e. volume per g) of a spherical macromolecule (radius R) then

$$u_2 = 4 M_2 v_2 / N_0 = \tfrac{16}{3} \pi R^3$$

and

$$B_{II} = 4 v_2 / M_2 = 16 \pi N_0 R^3 / 3 M_2^2 \qquad (2.32)$$

Other cases are more difficult to calculate, but Zimm (1946), for example, has shown that

$$B_{II} = L v_2 / M_2 d \qquad (2.33)$$

for rods of length L and diameter d, and that for very elongated ellipsoids of revolution

$$B_{II} = p v_2 / M_2, \qquad (2.34)$$

where p is the ratio of the long semi-axis of revolution of the ellipsoid to the short semi-axis. Clearly the excluded volumes and values of B_{II} will be much greater for elongated macromolecules than for spheres of the same molecular weight and specific volume. Since flexible chain molecules often behave as if they were inpenetrable spheres of an effective radius, ρ_{eff}, about 20 per cent less than that of their radius of gyration, ρ_g, the contribution of excluded volume to the second virial coefficient can be written, as for spheres, namely

$$B_{II} = 16 \pi N_0 \rho_{eff} / 3 M_2^2.$$

The ratio between the effective and true radii of gyration depends on the relation between the actual end-to-end distance of the ends of the flexible macromolecule and of a freely-rotating true random coil in which monomer segments are as likely to be found in contact with monomer segments as with solvent molecules. This constitutes the analogue in the statistics of flexible chains to the perfect gas since the contribution B_{II} of excluded volume effects to B is then zero: for the truly random coil in an ideal solvent this has the further meaning that ρ_{eff} is zero.

Biological macromolecules, even when single chains and not cross-linked to others, very rarely adopt the randomly-coiled configuration.† However, Tanford et al. (1967) have shown that in guanidinium chloride solutions of concentration 5 M or higher, many globular proteins, including several enzymes, are denatured and adopt a randomly coiled configuration whose end-to-end distance relative to the corresponding distance in the ideal, unperturbed random polypeptide can be estimated by means of equations relating B_{II} to this ratio and to the intrinsic viscosity. In principle these equations are an elaboration of those above but with further refinements (see Tanford et al. 1967, for further refs. and also Morawetz 1965, Chap. II, III, IV). In any case, all flexible macromolecules have much larger excluded volumes and values of B_{II} than spherical macromolecules of the same molecular weight and specific volume.

For many biological macromolecules the specific volume v_2, is in the range 0·6–0·8 so that B_{II} is of the order of $3/M_2$ (by eqn (2.32) if the molecule is spherical; but, by eqns (2.33, 34), v_2 is of the order $0·7p/M_2$ for elongated ellipsoid molecules, where p is the axial ratio (or L/d for a cylinder). Since, on the model of an ellipsoid of revolution, for many fibrous proteins and elongated molecules p can be of the order of 50 to 200 or 300, large variations in B_{II} might be expected, e.g. increasing by a factor of the order of 20 or more from spheres to elongated rigid molecules of the same molecular weight. The detailed equations suggest that flexible macromolecules come between these extremes. In fact, most osmotic pressure measurements have been made on biological macromolecules of molecular weights less than 200 000 and usually less than 100 000 since this is the easiest range in which to apply the method.

The biological macromolecules in this range of molecular weights are rarely highly extended and for most of the proteins studied the values of B_{II} (usually measured with Donnan effects suppressed, i.e. $B_{\text{I}} = 0$) have come within a factor of two of the theoretical value for a sphere of the same molecular weight and specific volume. Such proteins have indeed been what, on other grounds, are called *globular* proteins. Apart from Donnan effects (I), deviations from ideality of such solutions are more likely to be caused by the $\Delta \bar{H}_2$ term not being zero, or to the presence of dissociation or association equilibria (see next section). Myosin is one of the few fibrous proteins to be studied by osmometry: Portzehl (1950) showed by this means that BM_2 was 75 ml/g, which was about 25 times greater than that which the corresponding sphere would have had, implying an axial ratio of 128, if it is regarded as a rigid cylinder.

† See Flory (1953) who considers the interaction between overlapping macromolecules and mutual exclusion by spheres; and also Edmond and Ogston (1968) who consider the interactions in a mixture of fibrous threads and spheres.

Up to this point, the factors contributing to II in the second virial coefficient have been discussed assuming that the macromolecules were neutral. However as a macromolecule acquires a charge, the effective excluded volume is increased because one such molecule repels another like one so that the nearest distance of approach of their centres of gravity increases with charge. This increase of charge will tend to increase term II, as well as I, in the second virial coefficient. As with the Donnan effect, increase in the concentration of neutral salt (3) will decrease the effect of these repulsions, so as to eliminate the electrostatic factor in β_{22}, that is, in the 'excluded' volume. Equations for this electrostatic contribution have been derived by Scatchard et al. (1946).

III. The term III in eqn (1.51b) for the virial coefficient is $(-a_{23}^2/a_{33})$ and represents: (a) the interaction between macromolecule (2) and salt (3) since

$$a_{23} \equiv a_{32} = \left(\frac{\partial \ln a_3}{\partial m_2}\right)_{T,P,n_1,m} \quad (1.33)$$

and the binding coefficient of the salt on the macromolecule is given by

$$\Gamma_{32} = \left(\frac{\partial m_3}{\partial m_2}\right)_{\mu_3} = -\frac{a_{23}}{a_{33}}; \quad (1.43)$$

and (b) the non-ideality of the salt itself, since

$$a_{33} = \left(\frac{\partial \ln a_3}{\partial m_3}\right)_{T,P,n_1,m}. \quad (1.33)$$

The binding of salt represented by eqn (1.43) is the same phenomenon which shifts the minimum in plots of the virial coefficient against charge. For an 'ideal Donnan' osmotic equilibrium, in which the imbalance of the only two diffusible small ions is

$$\nu_{2+} = -\nu_{2-} = -z_2/2,$$

eqns (1.47, 48) show that this third term in the second virial coefficient may be written as

$$B_{\text{III}} = -\frac{a_{23}^2}{a_{33}} = -\frac{\beta_{23}^2 m_3}{2+\beta_{33}m_3}, \quad (2.35)$$

which is in a form whereby it may be usefully estimated. Lauffer (1966) writes this third term in the second virial coefficient as $2\xi_0$, where ξ_0 is the 'hydration factor', i.e. the weight in kg of solvent 'bound' by 1 mole of component 2, and is a function of β_{23}, β_{33} and β_{21}, when they are defined in terms of the anhydrous protein.

The relative magnitudes of the contributions, I, II, and III, to the second virial coefficient have been studied in detail in few investigations, the most extensive still being that of Scatchard et al. (1946) on bovine serum albumin solutions, the osmotic pressure of which was examined at various salt, as well as albumin concentrations. In this study, the interaction between the protein and salt (the chloride ions in particular) were exceptionally strong and III far from being ignored was utilized to determine the chloride binding. But for proteins the term III is usually relatively small, and of course negative, and only I and II have to be considered seriously. If one is working with the starred definition of component 2, term III automatically disappears from the expression (1.51*b*) for the second virial coefficient. So in most studies of proteins only I and II have to be considered. The relative magnitudes of these need to be assessed in each case; fortunately I can be derived approximately from the known ionic constitution of the mixture (*cf.* eqn 2.29) and can be compared with the observed virial coefficient to yield by subtraction the excluded volume effect II.

Apart from III, which is usually small for proteins, the other terms all lead to positive deviations from the ideal plot of π/c_2 versus c_2 (Fig. 2.3) with positive values of the second virial coefficient. In practice these deviations are not numerically great for most globular proteins in salt solutions of ionic strength 0·1 or greater since they amount to about only 2 to 4 per cent of RT/M_2 for 1 per cent (w/v, *i.e.* 10 mg/ml) solutions. In such cases the charge density is not high and it is no help to regard a proportion of the small counter-ions as actually 'bound' to or 'condensed' on the macro-ion. However, most synthetic polyelectrolytes, and some biological macromolecules (e.g. nucleic acids, proteoglycans, and glycosaminoglycans) contain an ionizable group on each constituent monomer unit. Macromolecules possessing such high charge densities deviate considerably from ideal behaviour and these deviations can be expressed either in the form of non-negligible, indeed highly significant, values for term III in the second virial coefficient, i.e. in values of activity coefficients far from unity; or in the form of a reduction in the net charge of the macromolecule, resulting from the 'condensation' upon it of small counter-ions. These theories have been developed principally in relation to studies on synthetic polyelectrolytes (Gross and Strauss 1966; Katchalsky, Alexandrowicz, and Kedem 1966; Manning 1969) but are beginning to be applied to nucleic acids and the components of model connective tissue systems (Meyer, Comper, and Preston 1971; Preston and Meyer 1971; Preston and Snowden 1972; Preston, Snowden, and Houghton 1972).

If the macromolecules of the same kind attract each other, the concept of an *excluded* volume appears to break down and, not surprisingly, negative deviations from the ideal line can be observed. Such attractions have, of course, to be strong enough to swamp the effects of electrostatic repulsion and are more usually described as macromolecular associations. This type of deviation from ideality is depicted in Fig. 2.3 and was, in fact, observed by Gutfreund (1948, 1952) with insulin in acid solvents, when the decrease of π/c_2 with increasing c_2, where π has already been corrected for Donnan effects, was attributed to increasing association of insulin subunits with total insulin concentration (see 4.4(a)). These observations and some more recent developments necessitate a closer examination of the effect of such association equilibria on osmotic pressure measurements, and in any case such associations are of immense interest in their own right.

2.8. Dissociating and associating systems

Since the osmotic pressure is a colligative property of a macromolecular solution it is sensitive to association and dissociation of the macromolecules, processes which affect the number of molecular species at a particular weight concentration. Many globular proteins, especially enzymes, contain a finite number of polypeptide chains held together by secondary forces (hydrogen bonds, electrostatic forces, 'hydrophobic' interactions) which are disrupted in the presence of guanidinium chloride, urea, detergents, and non-aqueous solvents, and sometimes also by variation in pH and ionic strength. If the conditions are such that the dissociation is reversible, then the degree of dissociation also depends on the protein concentration but sometimes solvent conditions can be found under which the protein is fully dissociated at all the concentrations studied. The solvent guanidinium chloride (5 to 8 M) is particularly effective for this purpose and its effectiveness may be enhanced by adding a reagent which breaks disulphide bonds, such as 0·1 M–0·5 M β-mercaptoethanol (Tanford *et al.*, 1966, 1967, 1968; Castellino and Barker 1968) or 0·1 M sodium sulphite (Tombs and Lowe, 1967). Urea (4 to 6M) had earlier been much favoured for this type of study (Edsall and Cohn 1943): such dissociation studies in these and other solvents have been reviewed by Kupke (1960, section IIIB) and have often been of critical importance in elucidating the structure of such compound proteins (e.g. arachin, conalbumin, insulin, keratins, silk fibroins, tropomysin).

The virtue of osmometry for this purpose is, that apart from the physical difficulties of handling solutions of such high concentrations of guanidinium salts, etc. which may lead to crystalline deposits and corrosion problems, the small molecule in the solvent can diffuse readily through the membrane and does not itself contribute to the osmotic pressure. Its disadvantage is that larger quantities of protein may be needed than, for example, in a sedimentation-equilibrium run but modern apparatus is overcoming this difficulty. Theoretically, its advantage is clear-cut when dissociation is complete at all the protein concentrations studied. For an extrapolation of π/c_2 against c_2 to zero c_2 eliminates non-ideality effects and yields at once M_n of the sub-units. Even if these sub-units are not all of the same size (*e.g.* the α and β chains in haemoglobin) the ratio of M_n in the absence of dissociating agent to that in its presence gives the total *number* of sub-unit chains present which are dissociable by the reagent in question.

This is a feature of any method which determines M_n and is particularly useful when the degradation of a macromolecule by a physical or chemical agent or by an enzyme is being followed. For the ratio of the original value of M_n to its value at any stage of degradation is equal to the number of breaks, plus one, which have been introduced into an original molecule of molecular weight equal to the initial M_n, so that the calculation can be performed whether or not the original macromolecular system was monodisperse (Charlesby 1954; Peacocke and Preston 1960): and it is the number of breaks introduced which is the parameter of interest in the degradative process. Determination of the number of sub-units of a protein in a dissociating solvent is based on the same principle, which is simply that the osmotic pressure depends on the number of molecules per unit volume and so on the number per unit volume of chain *ends*, too. Formally, for the dissociation of an originally monodisperse protein of molecular weight $(M)_0 \equiv (M_n)_0$ into its sub-units, of which there are i of molecular weight, M_i, one can write for the original solution:

$$(M)_0 = (M_n)_0 = \frac{c_0(\text{g/ml})}{m_0(\text{moles/ml})} = \frac{c_0}{c_0/(M_n)_0}$$

In the fully dissociating solvent, since $\sum_i c_i = c_0$

$$(M_n)_d = \frac{\sum_i c_i}{\sum c_i/M_i} = \frac{c_0}{\sum_i im_0} = \frac{c_0}{m_0 \sum i} \qquad (2.36)$$

$\therefore \ (M_n)_0/(M_n)_d = \sum i = $ (Total no. of sub-units per original protein molecule, regardless of their individual M_i).

Under solvent conditions less stringent than those just mentioned many proteins show a tendency to associate and it would be expected on the basis just outlined that osmotic pressure measurements might afford valuable indications concerning association constants. The basic theoretical problem in applying this, and indeed any other method based on equilibrium solution properties, is to allow for the effects of nonideality which, like the molecular weight of a self-associating system, also change in magnitude with total macromolecular concentration. One simplification may be introduced at once, however. The activity coefficients, γ_J on the m-scale, of each macromolecular component J, in any self-association process may be written (cf. eqn (1.30a) omitting higher terms) as a function of the virial coefficient, B:

$$\ln \gamma_J = 2BM_J c \qquad (2.39)$$

Here c is the *total* concentration of all the macromolecules and B is the same for all macromolecules regardless of size. The general equation for the association of macromolecules P_1 to form an aggregate P_n containing n of the original molecules is

$$nP_1 \rightleftharpoons P_n, \quad n = 2, 3 \ldots \tag{2.40}$$

This is a discrete type of association in which only two molecular species are present, P_1 and P_n. The thermodynamic equilibrium constant relating the activities a of the species in this equilibrium is

$$K_n = \frac{a_n}{(a_1)^n} = \frac{m_n \gamma_n}{(m_1 \gamma_1)^n} = \frac{m_n \cdot \exp(2BM_n c)}{m_1^n \cdot \exp(2BM_1 cn)}$$

where subscripts 1, n now refer to species P_1 and P_n, respectively (and not to thermodynamic components or number-averages as elsewhere). But since the molecular weights of the species on the two sides of the chemical equilibrium are related by $nM_1 = M_n$, the activity coefficients cancel to yield

$$K_n = \frac{m_n}{m_1^n}, \tag{2.41}$$

so that molalities ($\rightarrow$ molarities at these dilutions) can be used to obtain true equilibrium constants, even when the solutions are non-ideal, provided eqn (2.39) holds. Fortunately, higher terms in that equation are rarely necessary at the concentrations normally employed with solutions of biological macromolecules, so that the consequent simple form (2.41) can be used. An exactly parallel result is obtained if the c-scale is used† but the

† A derivation, which is more general than that given in the text, and is in terms of the c-scale, follows. Chemical reactions may be represented generally (Guggenheim 1967, § 1.44) as

$$0 = \sum_J \nu_J J \tag{i}$$

following the usual convention, with products set on the right hand side and ν_J, the number of moles of each chemical species J, having a positive sign if a product and negative if a reactant. If the system is not thermodynamically ideal, activity coefficients, y_J, on the c-scale may be written (cf. eqn 1.30a) as

$$\ln y_J = 2BM_J c \tag{ii}$$

where $c = \sum_J c_J$ is the *total* concentration and B is the same for all macromolecules regardless of size. Then, the thermodynamic true equilibrium constant for this reaction is given by

$$K = \prod_J a_J^{\nu_J}$$

Hence

$$\ln K = \sum_J \nu_J \ln c_J + \sum_J \nu_J \ln y_J$$

from (ii). Since in the chemical equilibrium mass is conserved, so $\sum_J \nu_J M_J = 0$, it follows that

$$\ln K = \sum_J \nu_J \ln c_J \tag{iii}$$

and

$$K = \prod_J c_J^{\nu_J}, \tag{iv}$$

which is the usual form for the concentration equilibrium constant, even though the solution is non-ideal. This simplification depends on the activity coefficients obeying eqn (ii).

molarity scale is more useful at the initial stages of the present discussion of osmotic pressures of reacting systems.

The self-association equilibrium to be considered can be written as

$$2P_1 \rightleftharpoons P_2;\ 3P_1 \rightleftharpoons P_3;\ \ldots iP_1 \rightleftharpoons P_i;\ \ldots jP_1 \rightleftharpoons P_j. \tag{2.42}$$

Here P_1, P_2, P_i etc., are 'monomer', (really the original associating macromolecule), dimer, i-mer, up to the limiting j-mer. On the basis that the solutions were *ideal* ($B = 0$), Steiner (1954) was able to formulate an important relation between the number fraction (z_1) of monomer present in the system at any concentration m and the number average molecular weight observed at that concentration M_{nc}, where the extra c in the subscript is meant to indicate that M_n is now itself a function of the concentration, even if $B = 0$. The relation may be derived as follows. If m is the total molarity, as mol/ml for convenience in this context, and c is the total concentration of macromolecules, as g/ml, then

$$m = \frac{c}{M_{nc}} = \sum_i m_i = \sum_i K_i m_1^i$$

$$= m_1 + K_2 m_1^2 + \ldots K_i m_1^i \tag{2.43}$$

by eqns such as (2.41). Hence

and so

$$dm = dm_1 + 2K_2 m_1\, dm_1 + \ldots + iK_i m_1^{i-1}\, dm_1 + \ldots$$

$$d \ln m_1 = \frac{dm_1}{m_1} = dm/(m_1 + 2K_2 m_1^2 + \ldots + iK_i m_1^i + \ldots)$$

But

$$c = M_1 m_1 + M_2 m_2 + \ldots M_i m_i + \ldots$$

$$= M_1 m + 2M_1 K_2 m_1^2 + \ldots iM_1 K_i m_1^i$$

by eqn (2.40). Substituting this in the previous equations gives

$$d \ln m_1 = \frac{dm}{c/M_1},$$

which on combining with $m = c/M_{nc}$ and eliminating c gives

$$\frac{M_{nc}}{M_1} = \frac{d \ln m}{d \ln m_1}. \tag{2.44}$$

Since

$$z_1 = m_1/m$$

and

$$dm_1 = z_1\, dm + m\, dz$$

$$d \ln m / d \ln m_1 = \left(1 + \frac{d \ln z_1}{d \ln m}\right)^{-1}.$$

THEORY OF OSMOTIC PRESSURE

Substitution of this in (2.44) and integration gives

$$\ln z_1 = \int_0^m \left(\frac{M_1}{M_{nc}} - 1\right) d\ln m, \quad \boxed{\text{Ideal}} \qquad (2.45)$$

which is the equation of Steiner (1954). For ideal systems $m = c/M_{nc}$, and is therefore obtained by the osmotic experiment so that z_1 is obtainable by a graphical integration of Steiner's equation. (M_1 is obtained by observing the value of M_{nc} as $c \to 0$ or from measurements in a dissociating solvent). Hence $m_1(=z, m)$ can also be calculated.

Inspection of eqn (2.43) shows that

$$\frac{d}{dm_1}[(m-m_1)/m_1] = K_2, \quad \boxed{\text{Ideal}} \qquad (2.46a)$$

and

$$\frac{d}{dm_1}\left[\frac{m-m_1-K_2 m_1^2}{m_1^2}\right] = K_3, \quad \boxed{\text{Ideal}} \qquad (2.46b)$$

so that successive K's can then be obtained from a knowledge of m_1 and m at various total concentrations.

If the process is a simple dimerization

$$2P_1 \rightleftharpoons P_2, \quad K_2 = c_2/c_1^2$$

and the solutions are ideal, then the above equations can be derived (q.v., for example, Nichol et al., 1964, eqn (7)) in the form

$$K_2 = \frac{1}{c} \cdot \frac{2M_{nc}(M_{nc}-M_1)}{(2M_1-M_{nc})^2} \quad \boxed{\text{Ideal}} \qquad (2.47)$$

and a plot of the r.h.s., without the c^{-1} term, against c should give a straight line of slope K_2. If the association process goes to an n-mer, eqn (2.40), then the general relation, if the solutions are ideal, is

$$K_n = c^{1-n} \cdot \frac{(n-1)^{n-1} n(1-M_1/M_{nc})}{(nM_1/M_{nc}-1)^n} \quad \boxed{\text{Ideal}} \qquad (2.48)$$

In this case, n has to be chosen so that a plot of the r.h.s. (without the c^{1-n} term) against c^{n-1} is linear.

When the solution containing self-associating macromolecules is *not ideal* ($B \neq 0$), although the equation (2.41) relating thermodynamic equilibrium constants to concentrations applies, the expression (2.45) relating the number fraction of monomers to the determined M_{nc}^{app}, does not, since M_{nc}^{app} is then given by

$$\frac{\pi}{cRT} = \frac{1}{M_{nc}^{app}} = \frac{1}{M_{nc}} + Bc = \frac{1+BM_{nc} \cdot c}{M_{nc}} \qquad (2.49)$$

(Compare, for example, eqn (2.4a) with terms higher than c^2 omitted). The combined effect of a term Bc increasing with concentration and of M_{nc} increasing through increasing aggregation, can lead to π/c vs. c plots showing a minimum. The following procedure has been suggested by Adams (1965a, b) as applicable to situations such as those expressed by eqn (2.49) in which both M_{nc} and Bc change with c and are both unknown.

The experimental observations (see Fig. 2.3) are the values of M_{nc} at different total concentrations c (dry wt. of macromolecule/ml of solution) and the molecular weight M_1 of the associating macromolecules, which may be called either a 'monomer', if the biologically active macromolecule is one covalently-linked chain, or a 'sub-unit', if several such chains constitute the active molecule. M_1 may have been obtained from the experimental values of M_{nc} extrapolated to infinite dilution, or from sequence or end-group studies or from studies with dissociating solvents. It has been shown (Adams 1965a, b, 1967; Adams and Fujita, 1963) that *for associating systems* (only) of either the discrete type of eqn (2.40) or of the more general type of eqn (2.42), the following relationships hold:

$$\frac{M_1}{M_{wc}} = \frac{M_1}{M_{nc}} + \frac{d}{d \ln c}\left(\frac{M_1}{M_{nc}}\right), \quad \boxed{\text{Ideal}} \quad (2.50a)$$

$$\frac{M_1}{M_{wc}^{app}} = \frac{M_1}{M_{nc}} + \frac{d}{d \ln c}\left(\frac{M_1}{M_{nc}^{app}}\right), \quad \boxed{\text{Non-ideal}} \quad (2.50b)$$

where M_{wc}^{app} is an apparent weight average molecular weight determined at concentration c, if non-ideal, or simply the actual weight average M_{wc} if ideal at concentration c.

Thus the values of M_{wc}^{app} for the associating systems at a given total c can be derived from the osmotic pressure results, which yield M_{nc}^{app}, by determining the slopes of the plots of M_1/M_{nc}^{app} against ln c. With this information, procedures have been described by Adams (during 1963–8) which give first B (or $2BM_1$), then c_1 and K_2 and then higher association constants. One important and useful relation which utilizes directly the observed M_{nc}^{app} is

$$\lim_{c \to 0} \frac{d}{dc}(M_1/M_{nc}^{app}) = (2BM_1 - K_2)/2. \quad (2.51)$$

So that if the solution is ideal ($B = 0$), K_2 can be obtained at once, even if higher associations occur (*cf.*, the procedures for using eqn 2.47). If no dimer is formed ($K_2 = 0$), then this limiting slope should give at once the non-ideality term $2BM_1$, and so B. However both determinations employing eqn (2.51) depend on a difficult determination of a limiting slope in the region ($c \to 0$) where experimental precision is least.

With M_{nc}^{app} at various c available from the osmotic pressure measurements, with M_1 known, and with M_{wc}^{app} calculated from eqn (2.50b), the general procedure, for equilibria of the type (2.42), which is advocated by Adams involves calculating: (a) an apparent weight fraction (f_1^{app}) of monomer (where the true weight fraction is c_1/c, and $c = \sum_i c_i$) which is given by

$$\ln f_1^{app} = \int_0^c \left(\frac{M_1}{M_{wc}^{app}} - 1\right) d\ln c$$

$$= f_1 \exp(BM_1 c), \qquad (2.52a)$$

(b)
$$M_1^2 \sum_i (c_i/M_i^2)^{app} = \int_0^c \frac{M_1^2\, dc}{M_{nc} M_{wc}}, \qquad (2.52b)$$

and (c)

$$\psi = -\left(\sum_i c_i M_i^2\right)^{app}/M_1^2$$

$$= \frac{d}{dc}\left(\frac{M_1}{cM_{wc}^{app}}\right) \Big/ \left[\frac{M_1}{cM_{wc}^{app}} - BM_1\right]^2, \qquad (2.52c)$$

all of which may be calculated from M_{nc}^{app}, M_{wc}^{app}, M_1 and c (although calculation of ψ rests on that of BM_1, which involves some successive approximation procedures). It cannot usually be known before making the analysis whether the type of association is like (2.40) or (2.42) and, if the latter, whether it is discrete, involving only particular association complexes, or an indefinite association with no limiting j. If it is discrete, the type of association can be ascertained by comparing the consistency and fit of various functions calculated in different ways (see Adams 1967) and by eliminating models which lead to zero or negative values of B, when the existence of a minimum in the plot of M_1/M_{wc}^{app} vs. c shows directly that B must be real and positive. If the association is indefinite, that is without a limiting value of j in eqn (2.42), the system can still be analysed, provided it can be assumed that successive equilibrium constants on a molar scale are equal, that is, that there is only one intrinsic association constant of the macromolecular monomer (P) to form all association complexes ($P_2, \ldots P_i \ldots$) The association of β-lactoglobin in 0·2 M acetate buffer (pH 4·6) at 16° appears to be such a case (Adams and Lewis 1968).

Derechin (1968, 1969a, b) has developed a different procedure, based on the multinomial theorem, and which also allows the calculation of equilibrium constants (K_1 to K_4) and of the virial coefficient B, without employing iterative procedures. It is equally valid for discrete and for indefinite associations, but involves knowledge of the type of population present in the system (e.g. state of highest aggregation or absence of a particular species). These procedures of Steiner, Adams, Derechin, *et al.* (but see Section 4.4(g)) have not been extensively applied to osmotic pressure observations, though a comparative study has been made of different types of the parallel computations of sedimentation equilibrium observations (Visser *et al.* 1972).

The theory has been extended (Steiner 1968, 1970; Laurent and Ogston 1963; Preston, Davis, and Ogston 1965) to systems in which two or more different macromolecules interact with each other. Steiner's treatment applies to colligative methods, such as osmotic pressure measurements which yield number-average molecular weights and, so far, is restricted to ideal systems. Systems of different interacting macromolecules are of considerable biological importance, e.g. the interactions of antigen and antibody, of hyaluronic acid and serum albumin, of haemoglobin and haptoglobin, of non-identical sub-units of enzymes, and of trypsin and chymotrypsin and other enzymes with protein inhibitors (see Section 4.4 for examples). There are very few methods which can give quantitative information on such systems and osmotic pressure measurements have the distinct advantage of not overweighting the higher complexes, since it is the lower complexes which are frequently of primary interest.

In practice the limitation of this, and other treatments, to ideal solutions is not a severe disadvantage since, as already calculated, the errors in an M_n calculated without allowing for non-ideality are rarely more than a few per cent for most globular proteins at concentrations of less than 10 per cent in solvents of ionic strengths greater than 0·1. This error is of the same order of magnitude as that in the experimental measurement itself. (There are a few small proteins which are very highly charged, such as histones, and with these non-ideality cannot be ignored (Haydon and Peacocke, 1968*a*, *b*; Diggle and Peacocke, 1968, section 4.4(g)).

Since most biological macromolecules, even in associating systems, are charged it is necessary to make the osmotic pressure measurements on solutions in the presence of a third salt component for the reasons already described (section 2.4). It has been shown (Adams 1965*b*) that one can apply the procedures just outlined to such systems provided all the macromolecular components are defined in the way recommended by Casassa and Eisenberg, i.e. as the 'starred' components of the account of Chapter 1.

2.9. Imperfect membranes and volume changes in elastic cells

The perfect semi-permeable membrane implicit in the general theory of osmotic pressure is, strictly speaking, a theoretical fiction. Real membranes have a finite thickness, the two faces may differ and of course it is impossible to be certain that some few macromolecular solute molecules cannot penetrate them. Fortunately, membranes used in practice possess sufficiently adequate semi-permeability, within the limits of experimental sensitivity. Ways in which they may be tested will be discussed in Chapter 3. The problems associated with partial permeability are more obvious in the case of many biological membranes and have received more attention in this context.

Clearly, if a macromolecular solute component, presumed to be impermeable, is in fact leaking through a membrane there can be no osmotic equilibrium at any time. Before the concentration of permeating solutes becomes equal on both sides of the membrane, there may be considerable transient flow of solvent and consequent pressure effects, so that it is more appropriate to apply the thermodynamics of irreversible, but microscopically reversible, processes (Katchalski and Spangler 1968). Staverman (1951) showed that, if π_m is the measured osmotic pressure, then

$$\pi_m = \sigma \pi_{\text{true}} \tag{2.53}$$

where σ is the 'reflection coefficient' and π_{true} is the true equilibrium pressure. It should be noted that a procedure such as extrapolation of π_m to zero time will not yield the true equilibrium value, because the solution will still be in contact with a leaky membrane. A further extrapolation to zero rate of leakage is required and the rate of leakage will be greatest at zero time because the concentration gradient will then be greatest. Extrapolation to zero concentration will yield an equilibrium value, so that if a series of osmotic pressures (π_m) are measured at different initial concentrations and are corrected by extrapolation to zero time, and then π_m/c_2 is extrapolated to zero concentration, the requirement of zero rate of leakage should be met. However, it is unlikely that a plot of π/c_2 against c_2 will be linear, so such a method is unsatisfactory—but fortunately rarely required in practice. The reflection coefficient may be used to characterize a membrane with respect to a particular solute and if it is determined, for example by comparison with a non-permeable membrane, it can be used to correct π_m. It has found little application in practice, and it is unlikely that σ would be independent of c_2 at high values of c_2. Where the objective is to study the mechanism by which solutes penetrate membranes σ is interpreted in terms of frictional coefficients; for example, it varies systematically when the solute and the membrane are hydrophilic or hydrophobic, but such studies are outside our scope.

The complications caused by leaking membranes in biological systems have been treated in terms of the reflection coefficient by Johnson and Wilson (1967) who used these ideas to develop relations describing the rate of volume change which results from transport of solvent through leaky membranes into cells. They find

$$\frac{dV}{dt} = P_w A \left(\frac{C_0 V_0 + S}{V} - C_s - C_m \right) \tag{2.54a}$$

and

$$\frac{dS}{dt} = PA(C_s - S/V), \tag{2.54b}$$

where V is the volume of the cell, P_w is a permeability coefficient, A is the membrane area, C_0 is the initial concentration (g ml^{-1}) of non-permeable solute in the initial volume V_0, S is the mass of permeable solute inside the cell at time t, C_s is its concentration in the external medium, C_m is the external concentration of the presumed non-permeable solute at time t, which in most cases will equal C_0.

These expressions do not allow for leaks in the membrane with respect to component C_m. Inserting the coefficient σ yields

$$\frac{dV}{dt} = P_w A \left(\frac{C_0 V_0 + \sigma S}{V} - \sigma \bar{C}_s - C_m \right) \qquad (2.55a)$$

$$\frac{dS}{dt} = PA(C_s - S/V) + (1-\sigma)\bar{C}_s \cdot \frac{dV}{dt} \qquad (2.55b)$$

where $\bar{C}_s$ is the mean solute concentration across the membrane. These relations correctly predict a rise in hydrostatic pressure to a maximum, followed by a fall to zero. The value of σ can be calculated from the slope of a plot of dV/dt against t.

Olmstead (1966) gives a useful account of the shrinking and swelling of mammalian cells as a result of differences in osmotic pressure between the external medium and the interior of the cell. The result of a large amount of experimentation is best summarized by

$$V_{\text{cell}} = \underline{R} V_{H_2O} \frac{\pi_0}{\pi_e} + b \qquad (2.55c)$$

where V_{cell} is the volume of the cell, V_{H_2O} is the volume of available water in the cell under iso-osmotic conditions. π_0 is the iso-osmotic pressure (i.e. the external osmotic pressure at which the cell neither swells or shrinks), π_e is the osmotic pressure of the medium, and b is another volume, which represents the part of the cell volume which is not acting as a solvent (i.e. roughly that occupied by dissolved solutes). $\underline{R}$, which is not the gas constant, then represents all the factors which cause the cell to deviate from behaving in ideal solutions as a perfect osmotic membrane.

The effects of non-ideality in solution will also include, in addition to those already discussed, a contribution from the elastic behaviour of the cellular structures themselves. This contribution may be treated as a 'constant' which characterizes a particular type of cell and has some predictive value in calculating volume changes. Biological systems such as intact cells are very complex—and may, for example, contain mixtures of ions and macromolecules at high concentration and there may also be several 'osmotically distinct' compartments within a single cell. Hence no rigorous relationships can be devised and semi-empirical equations such as eqn 2.55c, which at least recognize that osmotic effects are considerable, are the best available.

2.10. Solutes on both sides of a semi-permeable membrane

Consider the situation where a macromolecular solute, component 2, is present on both sides, A and B, of a membrane, which is impermeable to it. For compartment A, the osmotic pressure resulting from the presence of 2 is, according to eqn (2.4a), given by

$$\pi_A = RT\left[\frac{c_A}{M_2} + Bc_A^2\right] \quad (2.56a)$$

and for compartment B, by

$$\pi_B = RT\left[\frac{c_B}{M_2} + Bc_B^2\right], \quad (2.56b)$$

where subscripts denote the compartments. The resultant osmotic pressure is then given by

$$\Delta\pi = \pi_A - \pi_B = RT\left[\frac{(c_A - c_B)}{M_2} + B(c_A^2 - c_B^2)\right] \quad (2.57)$$

Note that if $\Delta c = c_A - c_B$ then the square bracket is not the same as $\Delta c/M_2 + B(\Delta c)^2$. Eqn (2.57) becomes

$$\frac{\Delta\pi}{\Delta c} = RT\left[\frac{1}{M_2} + B(c_A + c_B)\right] \quad (2.58)$$

and thus

$$\frac{1}{RT}\cdot\left(\frac{\partial(\Delta\pi/\Delta c)}{\partial(c_A + c_B)}\right) = B. \quad (2.59)$$

For relatively simple solutions, for which a π/c_2 against c_2 plot is linear, this method of finding B has no particular advantage, since B is constant and independent of c_2. It might, however, be used for determining the slopes when π/c_2 against c_2 plots deviate markedly from linearity because of association–dissociation reactions. Such a procedure is likely to be laborious, and has not, to our knowledge, been applied, though there is no reason why it should not be effective.

3

OSMOMETERS AND OSMOTIC PRESSURE MEASUREMENTS

OSMOMETERS of many different designs have been described, since until recently every investigator constructed his own. Those described here have been chosen to illustrate general principles or because they would still be the best instrument for certain measurements, or for both reasons. Osmometers designed primarily for non-aqueous solvents have not been included unless they can also operate with water as solvent. Some emphasis is laid on those errors which are associated with the actual construction of osmometers, and which may be alleviated by suitable modifications. A discussion of the problems associated with the choice and preparation of membranes is deferred (3.9) until after this account of osmometry.

3.1. General consideration

To obtain useful information it is necessary to measure the osmotic pressure developed at a series of solute concentrations. Since in all procedures the results are extrapolated to zero concentration, measurements have to be made at low concentrations and so at low pressures. This is especially necessary with high molecular weight solutes, whose molecular weights may increase with the concentration, as a result of polymerization, making extrapolations more difficult.

All osmometers contain two compartments, which are separated by a membrane, and also arrangements for measuring the pressure in, or applying a controlled pressure to either or both compartments. Pressure, in this context, may be either positive or negative. (The word 'suction', for a negative pressure will not be used.) Solvent is usually placed in one compartment and solution in the other, although there is no reason why two solutions of differing concentrations should not be used (see 2.10).

No classification of osmometers is very useful, although two kinds of procedure may be broadly distinguished. In the *static method*, the system is allowed to reach equilibrium without application of any controlled external pressures, other than that of the natural atmospheric environment. At equilibrium a hydrostatic pressure will have been generated so that the pressure on one side of the membrane is higher than on the other; this is then measured, and after correction is the osmotic pressure. In the *dynamic method* the rate of approach to equilibrium is measured, often under controlled applied pressures, and the equilibrium position when the net flow is zero is determined by interpolation. Some osmometers may be used in both ways, but most are normally better suited to one procedure rather than the other. In practice dynamic measurements are invariably made, since only so can it be ascertained that equilibrium has in fact been reached.

The simplest type of osmometer is illustrated diagrammatically in Fig. 3.1. Solvent passes through the membrane so that a hydrostatic head h is developed, which is constant when it becomes equal to the osmotic pressure.

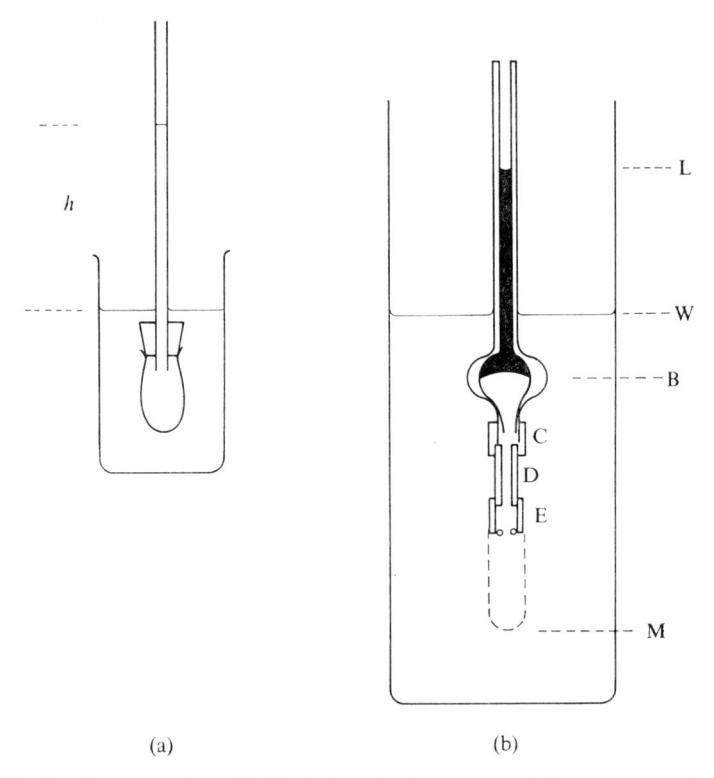

FIG. 3.1. Simple osmometers. (a) illustrates the simplest possible arrangement while (b) shows practical apparatus (after Adair 1961) using toluene as a manometer fluid.

The membrane must be adequate to support the full head of pressure developed. The disadvantages of this type of apparatus are numerous. The system is slow to equilibrate for it needs days to do so, during which time there is a serious risk of bacterial or fungal attack on biological materials. The concentration of the solute changes considerably during the experiment and is thus not precisely controllable. The height h can only be converted to pressures if the density of the solution is known, so this must also be determined. Because of the lengthy equilibration time required, membranes may alter their properties during the determination and minor leaks may become more significant; and for the same reason a separate osmometer has to be set up for each macromolecular concentration chosen which increases the quantity of macromolecule needed. In spite of all these problems, osmometers of this type have been and are still being used successfully and indeed, they may still be the method of choice for very low osmotic pressures. A striking demonstration of this is the remarkable results, obtained by Adair (4.1 p. 106).

Figure 3.3 shows a plot of the approach to equilibrium in a simple osmometer in which the rate of change of pressure decreases to zero as equilibrium is approached. Such a plot is also indicating the rate of solvent transport through the membrane. If, therefore, we could use the rate of solvent transport as a measure of the pressure difference across the membrane, by choosing suitable applied pressures we might be able to extrapolate or interpolate to find the applied pressure at which there is no net solvent transport. It has been found that it requires only a few minutes, with most membranes, for a steady rate of flow to be established under a constant pressure head.

This last mentioned approach is the basis of more recent osmometers, and is the basis of their relatively rapid operation. In earlier types, the pressure head was developed in the osmometer itself, but later an external manometer to apply controlled pressures to either compartment was naturally developed. This is now a quite standard practice. Solvent flow through the membrane is measured either against some datum mark in a capillary or by pressure-sensing electronic devices such as transducers.

If the rate of solvent transfer across the membrane is dv_1/dt under an applied pressure π_1 and dv_2/dt under pressure π_2 then

$$\pi + \pi_1 = K\frac{dv_1}{dt} \quad (3.1)$$

$$\pi + \pi_2 = K\frac{dv_2}{dt} \quad (3.2)$$

and this can be solved for the unknown osmotic pressure, π. However, the relationship between pressure and solvent flow may not have this simple linear form, and K may not be constant. In actual practice, linear relationships seem often to hold (e.g. Fig. 3.3 and 3.6) but this must be established for each osmometer and for each membrane used.

There are no serious technical problems in measuring differences in manometer levels to the required accuracies: manometers equipped with cathetometers are capable of measurements of 100 μm of water pressure, and optical devices may be used instead of cathetometers to determine levels down to better than 0·5 μm. In electronic equipment, transducers with suitable amplifiers are available to measure pressure changes of less than 10 μm. Most measurements fall in the range 10–1000 mm of water. In general, the precision of pressure measurements, (though not necessarily of osmotic pressures) is well above that of the determination of concentration of biological materials, and is not a source of major error.

It is emphasized that the measured pressure is not necessarily the osmotic pressure because surface tension effects leading to capillary rise must always be present and must be corrected for. The use of wide tubes helps to reduce these effects, though it may increase the volume required, but they are not negligible and even for 1 cm diameter tubes may amount to 0·5 mm H_2O. The difficulty of accurately estimating this correction means that the precision of *osmotic* pressure measurements is rather less than that of simply measuring the pressure exerted by a manometer.

Temperature control is necessary not only because of the need for stabilizing pressures in most designs of osmometers, but also because the osmotic pressure is directly proportional to the absolute temperature: a change of one Celsius degree thereby produces a change of about 0·33 per cent in the value of π, at room temperature. Unfortunately most osmometers contain a closed chamber, in which the pressure is measured, and thermal expansion or contraction of the solvent then causes large fluctuations, so that the instrument becomes a thermometer. These fluctuations are transient in the sense that solvent flow through the membrane eventually leads to loss of the pressure generated, but this can be a relatively slow process. The need for a constant temperature should however, be distinguished from the need for a precisely known temperature. Generally, osmometers need a constant temperature (to $\pm 0.001°C$) for periods of a few minutes, but it is not often necessary to know its absolute value to better than $\pm 0.05°C$.

All dynamic osmometers measure rates of pressure change in a vessel of some sort, and to obtain maximum sensitivity it follows that except at the point of measurement (e.g. a transducer or a capillary meniscus) the walls of the vessel must be rigid. However, one wall must always be the membrane, which is inevitably to some extent flexible. Various arrangements have been used to clamp it, so that it moves as little as possible but this is a source of many difficulties in making rate measurements and is always an important part of osmometer design considerations. Much the most significant change in recent years is the adoption of instruments which measure rates of solvent flow and the discovery that this reaches a steady state quickly. It is now possible to make measurements in a day which formerly required three months and this, coupled with careful design to minimize volumes of solution has eliminated some, if not all of the practical difficulties previously associated with this technique.

Some actual osmometers will now be discussed in detail, in the light of the foregoing.

3.2. Simple osmometers

The first precise measurements of the osmotic pressures of protein solutions were made by Adair (1925) (1961) by using very simple osmometers. It is doubtful if these osmometers would often be chosen nowadays, but they are effective and may still find some application. Figure 3.1b illustrates one type in which the membrane is in the form of a bag that is filled with solution over which toluene is layered. The tubes C and E are of rubber and D of glass, to facilitate attachment of the membrane. As a result of osmosis the toluene rises up the capillary to some level L. The bulb at B provides a reservoir of toluene, so that solution does not reach the capillary.

The levels at L, the diffusate level W, and the level of the toluene–solution interface B are then determined relative to a level at the base of the membrane. Then the developed pressure must be

$$(L-B)\rho_t + (B-M)\rho_s - (W-M)\rho_d$$

Where ρ_t is the density of toluene, ρ_s that of solution and ρ_d that of diffusate. This must be corrected for capillarity effects, and with this design there are three interfaces—L, W, and B. The simplest method of correction is to remove the membrane, and fill the solution compartment with solvent. The level W may be adjusted so that B and L are near the levels used in measuring the osmotic pressure. This will then yield a set of values for the levels of L_2, B_2, and W_2. The capillary correction (C) is then $C = (L_2-B_2)\rho_t + B_2\rho_d - W_2\rho_d$. and the osmotic pressure is obtained as:

$$\pi = (L-B)\rho_t + (B-M)\rho_s - (W-M)\rho d - (L_2-B_2)\rho_t + B_2\rho_d - W_2\rho d. \quad (3.3)$$

The capillary correction is small but not negligible, and it might be objected that it is not precise because solvent, and not solution, is used in determining it. However, the error so introduced is unimportant in most circumstances. It can be determined by further experiment if necessary, by comparing the capillary rises for solution and diffusate. Further correction for the column of air of height $L-W$ may be made, but this is negligible unless very long columns of toluene are involved, since air has a density of only $1 \cdot 18 \times 10^{-3}$ g/cm^3.

In a simpler method, the toluene is omitted and the capillary rise of the solution is recorded. The osmotic pressure is then given by

$$\pi = (L-M)_{\rho s} - (W-M)_{\rho d} - C. \quad (3.4)$$

Since ρ_s and ρ_d are often not significantly different, within the overall precision of the method, this can be further simplified to

$$(L-W)_{\rho d} - C = \pi \quad (3.5)$$

This type of osmometer requires several days to come to equilibrium, and its use has to be preceded by dialysis of the solutions over at least 48 hours, when bacterial attack can occur. Adair (1961) recommends the use of a room maintained at 1°C. It is also fair to say that the membranes are difficult to make and to attach to the apparatus. In particular they must support the full hydrostatic head developed, and over the long periods required for equilibration they may stretch, change their permeability characteristics, and leak slightly. The detailed descriptions given by Adair (1925, 1928) give a good indication of the high level of technical expertise required to obtain consistent results. Adair (1961) has described more highly developed though still fundamentally similar versions of this type of osmometer.

3.3. Osmometers with clamped flat membranes

The osmometer introduced in 1943 by Fuoss and Mead (Fig. 3.2) is of interest because it seems to have set the pattern of subsequent developments

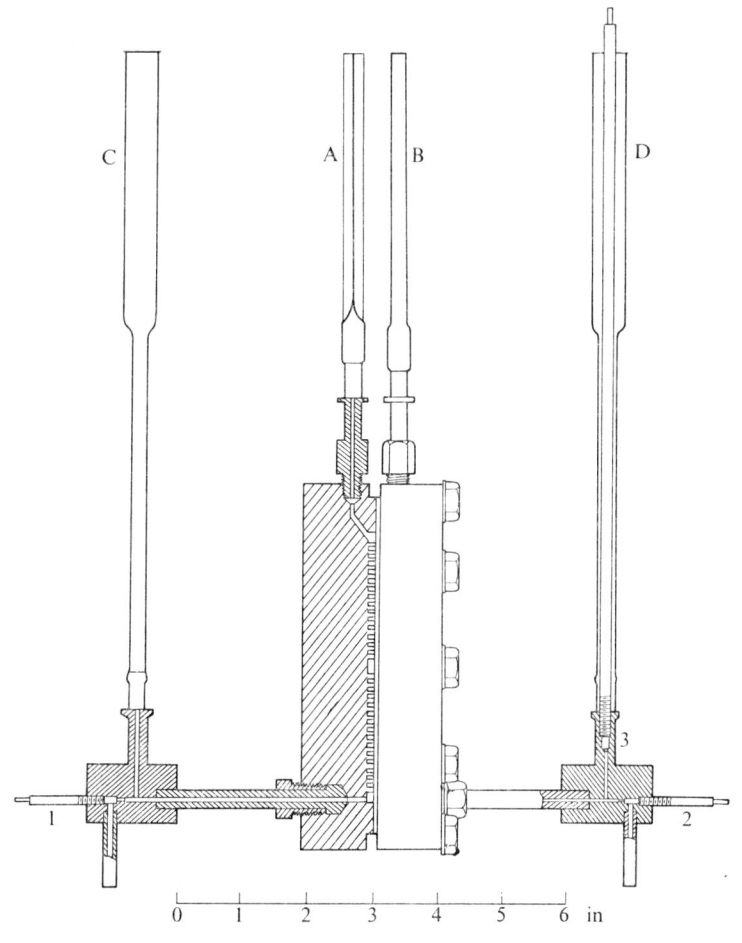

FIG. 3.2. The osmometer of Fuoss and Mead (1943). The membrane was clamped between stainless steel plates, grooved to permit access of solutions to the membrane. C and D are for filling the two sides, and can be sealed by needle valves 1 and 3. Solvent flow is observed in capillaries A and B.

with respect to methods of holding the membrane. Bag-shaped membranes are difficult to make and secure in the apparatus while sheets are much easier to make with controlled properties. Fuoss and Mead found that a flat membrane could be clamped between metal faces, suitably grooved. In the original version there was no provision for applying external pressures, and the osmotic pressure was determined by measuring the difference, suitably corrected for density differences between the levels of solvent and solution in the capillaries A and B. Any measurements of height in capillaries with a cathetometer requires that the capillary be exactly vertical if the height measurement is to be translated directly to pressures, and this is not easy to arrange, even when plumb lines are employed. These difficulties were avoided if the two columns A and B always tilted by the same amount: This was achieved in this osmometer by the rigid mounting of the capillaries.

Although this osmometer could be allowed to come to equilibrium, Fuoss and Mead made dynamic measurements. The level of solvent in the tube B was raised, and the rate of change of the level in capillary A was measured. Because B was 1·3 cm wide and A only 0·1 mm, A could vary considerably with no appreciable change in the level in B. The applied pressure was then effectively constant, and could be varied over a small range by varying the level in B. Fig. 3.3a shows that the rate of movement of the meniscus, and hence the rate of solvent flow through the membrane, was linearly related to the applied pressure. Fig. 3.3b shows a method for extrapolation to the equilibrium position: a pressure slightly above the osmotic pressure was established, and the position of the meniscus was followed. A pressure, as nearly as possible an equal amount below the osmotic pressure was then chosen, and the meniscus, which then moved in the opposite sense, was followed. The central line was the half sum of the two curves, and was a very close approximation to the asymptote. This method was less laborious than that shown in Fig. 3.3a and required only two sets of measurements. In both cases the pressure causing zero solvent flow corresponded to the osmotic pressure.

Since the pressure measurements are made by comparing the height in the two tubes A and B, a capillary correction is necessary. This may be combined with a general zero point test by placing solvent in both cells, with a membrane in position. It is said that some membranes produce an apparent osmotic pressure (with solvent on both sides of the membrane) even with corrections for capillarity taken into account. This is the so called membrane asymmetry effect; its direction is usually associated with one face of the membrane, and it may amount to several millimetres. Such effects must be due to contaminants or elastic effects in some part of the osmometer. If a membrane could really be found which could maintain a pressure difference it would be easy to construct a perpetual motion machine! In practice, it is a good idea before making any measurements, to check the zero positions with a

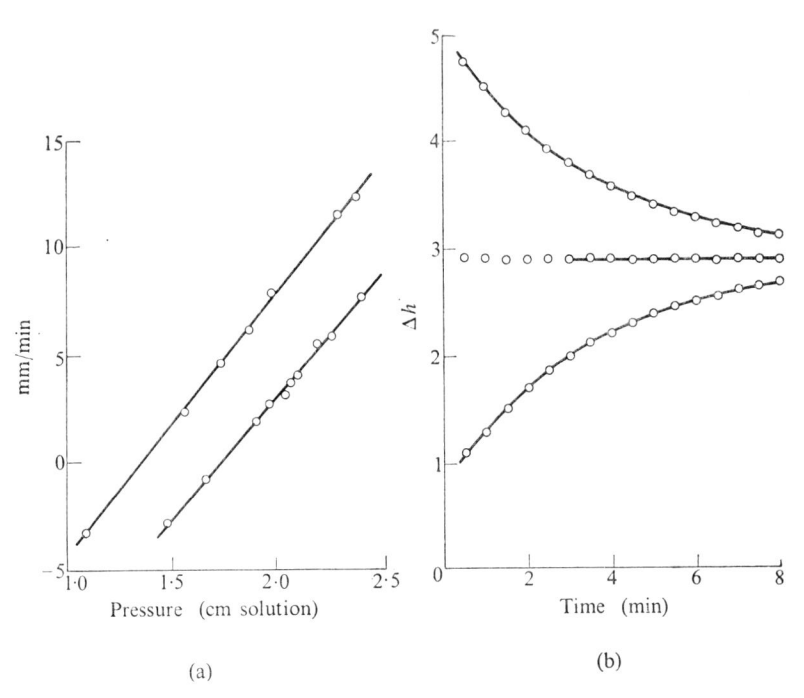

FIG. 3.3. Solvent transport through the membrane in a Fuoss-Mead osmometer (Fuoss and Mead 1943). (a) shows the rate of change of meniscus position in capillary A in Fig. 3.2 under various applied pressures. Note the linear relationship. (b) shows the actual position of the meniscus over a period of time, as equilibrium is approached from above, and below, the equilibrium position. The central line is the half sum of the two curves, and approximates to the mean of the asymptotes of the curves.

membrane in place and solvent in all compartments of the osmometer. There are no closed compartments in the Fuoss and Mead osmometer, and temperature fluctuations are not, therefore, particularly important. Fuoss and Mead generally measured osmotic pressures of only a few cms of water, but remarked that at higher pressures flexing of the membrane disturbed flow-rate measurements. This is a point of some importance in more elaborate osmometers designed more recently and will be mentioned again. It required only a few seconds for a steady state of flow to be set up and an osmotic pressure could be obtained 20 min after running solution into the machine. A comparison of this method, with the simple osmometer described in section 3.2 shows that the technique of Fuoss and Mead incorporated many of the essential points that characterize apparatuses in current use.

3.4. Dynamic osmometers with controlled external pressures

The Fuoss and Mead pattern is restricted to pressures of only a few cms of water because use is made of the capillaries A and B (Fig. 3.2) to apply a pressure. A natural development is to attach a manometer to either side, so that more easily controlled, and measured, positive or negative, pressures may be applied. A number of designs have been published (Masson and Melville 1949; Scatchard, Batchelder and Brown 1946) and that of Guntelberg and Linderstrom-Lang (1949) has been shown to be particularly precise.

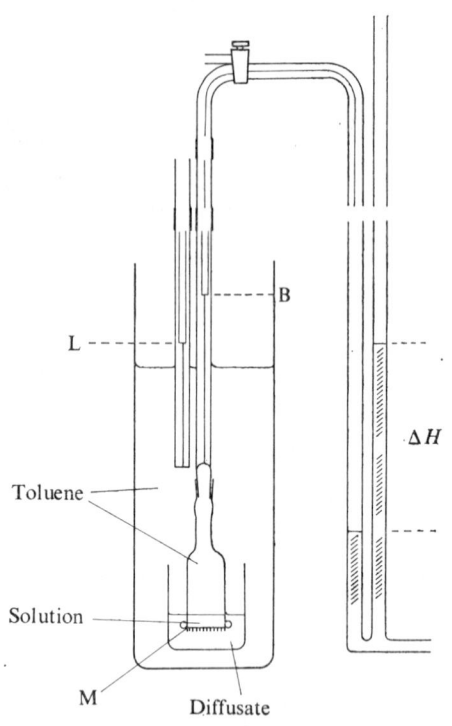

FIG. 3.4. The osmometer of Guntelberg and Linderstrøm–Lang (1949). The rate of change of the position of the meniscus B was observed under various pressure heads ΔH, giving plots similar to that shown in Fig. 3.3a.

The arrangement is shown in Fig. 3.4. The membrane at M is clamped to a perforated steel plate, which fits against a ground glass flange to form a seal. The solution is overlaid with toluene, so that the osmotic pressure produces a capillary rise to level B. The solvent is also covered with toluene, and an external manometer was fitted as shown: note particularly that there was an air gap between the manometer and B. The manometer may contain any suitable liquid: water was originally used in this case. L is a capillary, identical with B and provides a continuous capillary correction. The rate of movement of the meniscus at B was measured under various applied pressures ΔH, from the manometer. There was a linear relationship between rate of solvent transport, and applied pressure over the range used. The pressure for zero transport may be obtained by interpolation, and must then be corrected by subtraction of the quantity $(B-L)\rho_t$ where ρ_t is the density of toluene, which includes all necessary capillarity corrections. The instrument was thermostated by immersion in a water bath, fixed at $20\pm0\cdot003°$. A steady flow rate was obtained in five minutes, though 24 hours were needed for stabilization after the machine was set up for the first time. It has to be dismantled to change the solution, and a set of measurements at a series of concentrations required several weeks. Guntelberg and Linderstrom-Lang point out that the volumes transported through the membrane were of the order of microlitres so that the equilibrium at the membrane, and the composition of the solution at the membrane face changed very little. A recent analysis of dynamic osmometry (Davies 1966) suggests that this may have been the main reason for the success of this instrument. A criticism of the use of toluene in this and other instruments is that it interferes with ultraviolet absorption measurements on the solution, and that proteins tend to precipitate at the toluene–water interface.

An instrument which was generally related to the preceding and which attained high precision was that designed by O. Smithies (1953). It differs in using an unsupported bag-type membrane, which had to be specially formed. Also, solvent flow through the membrane was determined by observing the rate of movement of droplets of o-chlorotoluene dispersed in solvent, rather than at a meniscus. Of particular interest is Smithies' method of eliminating capillary effects and density differences by substituting for the semipermeable membrane a totally impermeable, but flexible, membrane. It was partly collapsed so that pressures could be transmitted through it by flexing, and it is easy to see that all the pressures and corrections necessary for the instrument could be measured in the complete absence of the osmotic pressure, which could not arise in this case. It might, however, in practice, be difficult to be certain that the membrane is completely impermeable to all solutes, and is completely flexible, and in any case the method cannot be employed in instruments where the membrane is clamped.

3.5. Osmometers with electronic pressure-sensing devices

It is convenient to consider these machines together, though they do not differ fundamentally from those described in the preceding section: nor are they an homogeneous group since they differ in their flexibility of use, and even in whether dynamic or equilibrium measurements are made.

Rowe and Abrahams (1957) described one of the earliest osmometers of this kind, which eventually became commercially available.† This instrument has been used, with slight modification, by others (Tombs and Lowe 1967) and a detailed study of its characteristics has been made (Davies 1966). The arrangement is shown in Fig. 3.5. In use the instrument is filled

† From Nash and Thompson Ltd. Chessington, Surrey.

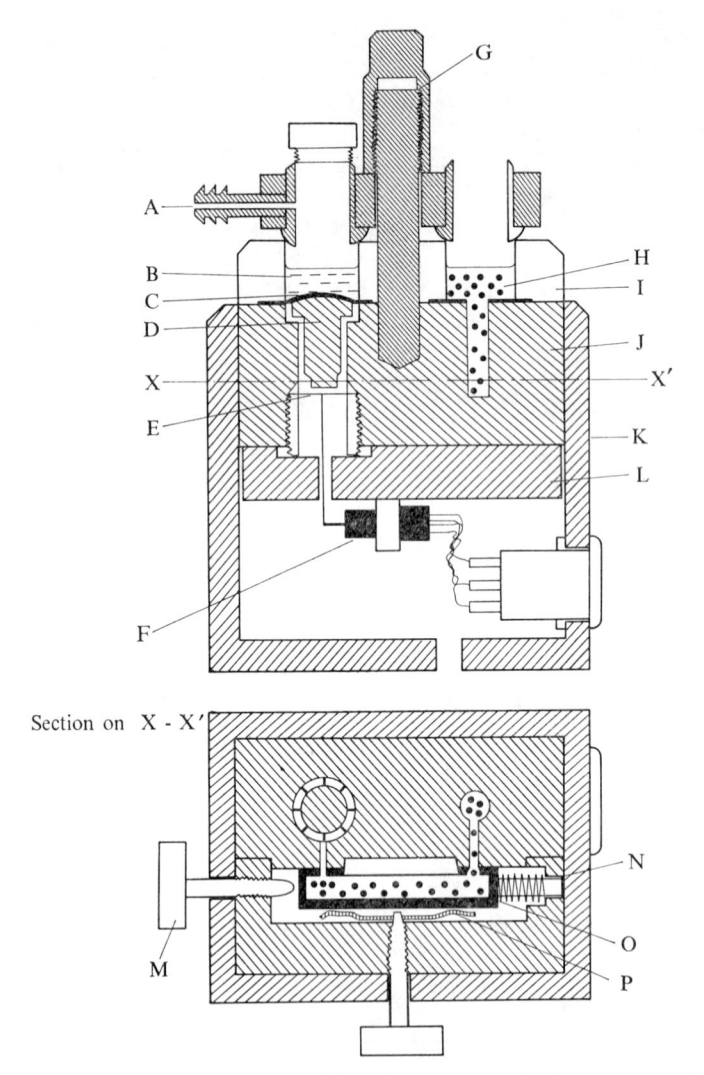

76　OSMOMETERS AND OSMOTIC PRESSURE MEASUREMENTS

FIG. 3.5. An electronic osmometer (taken from Rowe and Abrahams 1957). A. Pressure connection to an external manometer. B Solution compartment C membrane. D membrane support. E platinum diaphragm. F transducer. G clamp. H solvent compartment. I perspex block. J stainless steel block. K, L insulating plastic case. M, N, O and P tap components.

with solvent, and the dome-shaped membrane support *D* inserted. After suitable flushing to remove bubbles, a membrane is laid over the support, the perspex top *I*, lowered on to it, and finally clamped in place by a screw *G*. Solvent is put in at *B*, to cover the membrane: the solvent level at *H* is also adjusted to be nearly the same. The membrane is slightly stretched during clamping, and should be left for at least an hour to settle down. The tap is left open during this period. Pressures may be applied from a manometer, through A: as a preliminary check, a pressure of about 20 cm of water is applied. If there are any holes in the membrane, the levels at B and H will change visibly and rapidly. There will be a much slower change if the membrane is intact.

The instrument may now be zeroed: the levels in B and H are adjusted to coincide, and controlled pressures applied through A, over a range of -2 to $+2$ cm of water. The electrical circuit consists of a simple Wheatstone bridge, with a spot galvanometer, sensitive to movements of the transducer valve. The valve is actuated by a platinum membrane E, and is thus sensitive to pressure changes in the chamber below the membrane, which appear as arbitrary units on the galvanometer scale. The galvanometer is first zeroed by means of variable resistances, and a selected pressure applied for at least ten minutes: the tap is left open. It is then closed for a suitable time: Rowe and Abrahams recommend one minute, but up to four minutes causes no difficulty. While the tap is closed solvent moves through the membrane, and there is a pressure change, causing a deflection of the platinum membrane E. The electronics of the original machine were not sufficiently stable for this to be measured reliably. Instead, the sharp change when the pressure was released by opening the tap was used: the galvanometer kicks and the magnitude of the kick is plotted against the applied pressure. If the instrument is working correctly, zero applied pressure must correspond to zero deflection, and zero kick.

A typical curve (A) obtained when the membrane is properly clamped is shown in Fig. 3.6. The slope is a good measure of the sensitivity of the membrane, and with this one, for example, a precision of ± 0.1 mm of water was obtained. Curve B illustrates the effect of a common fault in which the sensitivity apparently depends on the direction of the applied pressures. This is due to flexing of the membrane and although the zero point is correctly indicated such membranes are best discarded. C is the result obtained with a membrane with poor solvent permeability: quite apart from the low sensitivity, the relationship between transport and pressure is no longer linear, although the zero point is still correctly indicated. In fact, line A also departs from linearity over a wider pressure range but this is unimportant because the actual pressures needed to make measurements can be restricted to the linear part of the range. The effect of a hole in the membrane is to cause an apparent increase in sensitivity to such an extent that the nearly horizontal line D is obtained.

Solvent is then removed from compartment B, and a solution inserted. About 0·5 ml is sufficient. Suitable pressures are applied, exactly as for the zero check, and a plot made, as in Fig. 3.6. Zero deflexion then corresponds to the osmotic pressure. In practice, large transient fluctuations are seen when solution is inserted, or changed back to solvent, but these disappear

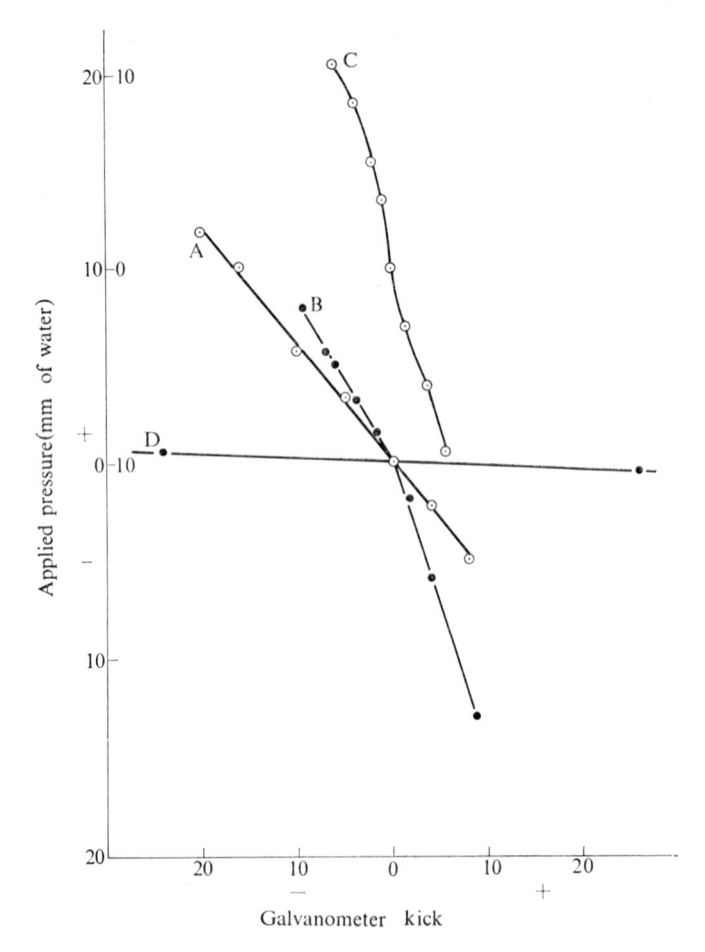

78   OSMOMETERS AND OSMOTIC PRESSURE MEASUREMENTS

FIG. 3.6. Solvent transport plots for the Rowe–Abrahams osmometer. The 'galvanometer kick' is a measure of the rate of pressure change at the platinum diaphragm E, in Fig. 3.5. A is the result for a good membrane properly inserted. C, B, and D correspond to various defects, and are explained in the text. For the sake of clarity the pressure scale for C has been shifted.

after a few minutes. The slope of the plots remains constant up to at least 300 mm of water, if the membrane has been properly clamped. When all the solutions have been measured, solvent is put in, and the zero check repeated. Any solute permeation of the membrane, or any change in the apparent permeability characteristics of the membrane are at once revealed by its failure to coincide with the initial zero check. A membrane may be used to made about fifty measurements before its sensitivity begins to decrease noticeably.

OSMOMETERS AND OSMOTIC PRESSURE MEASUREMENTS 79

A number of factors have been mentioned in the preceding discussions, such as rates of solvent transport, sensitivity, membrane flexing and so forth. Davies (1966) has attempted to derive a mathematical description of the various factors involved, and the account that follows is based on this treatment, and a similar approach made by Schulz and Kuhn (1961).

In an osmometer, with solvent in both compartments, let Δp be the pressure difference across the membrane, with a higher pressure in the solution compartment. Let the volume in the solvent compartment be V. The rate of change of pressure (that is, of increase of pressure in the solvent compartment by solvent transport through the membrane) can be described by

$$\frac{d\Delta p}{dt} = -k_1 k_2 \Delta p \tag{3.6}$$

while

$$\frac{dV}{dt} = k\,\Delta p \quad \text{and} \quad \frac{d\Delta p}{dV} = -k_2.$$

These are first order rate equations, and state for example, that the rate of change of volume is directly proportional to the pressure difference across the membrane. This is experimentally justified, for example by the curves obtained by Fuoss and Mead (Fig. 3.3). The permeability of the membrane is described by k_1, a flow constant, which may be further broken down to

$$k_1 = p \times A \tag{3.7}$$

where p is the 'permeability' in any convenient units (e.g. cc sec^{-1}/unit pressure) and A is the membrane area accessible to solvent.

Given a constant membrane (and hence constant k_1), k_2 is the factor that is of most interest and determines the sensitivity of the osmometer: it may be called the volume constant, for, when expressed as a pressure–volume ratio, it is the rate of volume change produced by a given pressure (e.g. cm of water/cm^3).

In the osmometers described in Section 3.3, a capillary rise or fall is observed which can be expressed in terms of capillary radius and is $1/\pi r^2$. There is obviously a lower unit for usable capillary sizes, say 0·2 mm diameter and the lowest observable rate of movement is of the order of 0·01 mm/s. The volume change observable is therefore $0{\cdot}01 \times (0{\cdot}1)^2 \times 10^{-3}$ cm^3/s i.e. about 3×10^{-9} cm^3/s.

The lowest measurable rate of pressure change is of the order of 0·01 mm/s.

$$\frac{\Delta p}{V} = \frac{10^{-3}}{3 \times 10^{-9}} \simeq 3 \times 10^5 \text{ cm/cm}^3$$

These are extreme limits, and a more realistic estimate would be 2 mm capillaries, and 0·1 mm/s for rates. A value of 10^3 cm^{-2} is probably the best actually achieved in practice. In the Rowe–Abrahams osmometer, the situation is more complicated. The effect of volume change is to flex the platinum diaphragm, and thus k_2 depends entirely on the diaphragm properties. The effect of flexing is further multiplied by the geometry of the transducer itself. The result, in practice, is to increase k_2 to about 10^4–10^6 cm^{-2} but because of the further complications of the elasticity of the diaphragm, eqn (3.6) may not be valid and may depart from first order kinetics. There is experimental evidence, that in the Rowe and Abrahams machine equation 3.6 is valid over most of the pressure range, while departures are observed at low pressures. As already pointed out this is allowed for in the experimental procedures and does not in any case render the results any less precise. It may vitiate theoretical discussions however.

It has been assumed so far, that any change in V will be entirely manifested in either a meniscus movement, or a flexing of the platinum membrane. This cannot be entirely true: the solvent itself is compressible, the walls of the cell must have some elasticity and in particular, the membrane may stretch under applied pressures. The bulk modulus of water is $2 \cdot 2 \times 10^9$ cm^{-2} (Davies 1966) and if k_2 is equal to or greater than this, the result of osmosis will be merely to compress the solvent slightly, with no effect on the pressure transmitted. If k_2 is of the order of 10^7, solvent compressibility effects will still be significant, so that k_2 values of the order of 10^6 are, in fact, at the upper practical limit for aqueous solvents: at this level, compressibility effects are about 0·1 per cent of the measured pressures and may be considered unimportant. It is reasonable to assume that the materials of the osmometer cell are less compressible than water and any errors from this cause must be entirely negligible. However, the effects of their elasticity may not be negligible, though the membrane is certainly more elastic (i.e. stretchable), than any other part.

Let this stretching produce a volume change ΔV.
Then
$$k_2 = -(a - \Delta V - c) \tag{3.8}$$
where a is dependent entirely on the geometry of the machine, and c represents solvent compressibility effects. Moreover,
$$\Delta V = b \Delta p, \tag{3.9}$$
since the extent of membrane stretching will depend on the magnitude of the pressure difference. This relationship is not valid over more than a small range of pressures, since the stretchability of membranes decreases sharply as they are stretched. Thus
$$k_2 = -(a - b \Delta p) \tag{3.10}$$
$$\frac{d \Delta p}{dt} = -a + b \Delta p \tag{3.11}$$
neglecting c.

It is unfortunately easy to set up an osmometer so that $b\,\Delta p$ is greater than a, with the result that the instrument has no sensitivity at all. It is also evident that the apparent sensitivity varies with the pressure. The equation

$$\frac{d\,\Delta p}{dt} = -PA(a - b\,\Delta p)\Delta p \tag{3.12}$$

thus describes more or less adequately the main factors arising from the shape and materials of the osmometer, and the size, permeability, and flexibility of the membrane.

The effect of temperature fluctuation is to produce a volume change inside the compartment. Its magnitude must be given by

$$V = -V\frac{dT}{dt}e \tag{3.13}$$

where e is a constant dependent on coefficients of expansion of the osmometer materials and solutions. Therefore

$$\frac{dV}{dt} = k_1\,\Delta p + v \tag{3.14}$$

and at an equilibrium, where $dV/dt = 0$,

$$\Delta p = -\frac{v}{k_1} = \Delta_{PT} \tag{3.15}$$

Hence

$$\frac{d\,\Delta p}{dt} = k_1 k_2 (\Delta p - \Delta_{PT}) \tag{3.16}$$

and Δ_{PT} can be seen as a constant zero-error: this will be the case however, only if dT/dt is constant. Again, if Δ_{PT} is large compared with Δp, measurement will in effect be of the temperature change, and not of the osmotic pressure. The instrument will in fact be in use as a thermometer. Any effect that produces a small volume or pressure change inside the osmometer will have the same result. Small trapped bubbles, for example, will produce a constant zero-error. Similar effects can arise if the membrane exerts a pressure: membranes are usually stretched slightly during insertion, to remove flexing effects. This may result in their exerting a constant mechanical strain: fortunately this is slowly lost, and is one reason why an osmometer takes several hours to 'settle down' after initial assembly.

If there are leaks in the osmometer (and not necessarily in the membrane) the rate of leakage may be described in terms of a flow constant k_3, the net pressure at the point of leakage $H - \Delta P$ and a measure of the magnitude of the leak, Δ_{PL}.

If w is the rate of volume change, when $d\Delta p/dt = 0$

$$w = -k_3(\Delta_{PL}+H-\Delta p) \tag{3.17}$$

from which

$$\frac{d\Delta p}{dt} = -k_2(k_1+k_3)\Delta p - k_2 k_3(\Delta_{PL}+H) \tag{3.18}$$

At equilibrium the pressure will continue to change by Δ_{PL}, so that

$$\Delta_{PL} = k_3 H/k_1 \tag{3.19}$$

and the zero error is always positive. As with all the other factors considered, if Δ_{PL} is large compared with Δp, osmotic pressures cannot be measured.

Zero errors frequently occur in osmometry: they are usually very small compared with the pressure to be measured, but any kind of unexplained error is disquieting. The value of this treatment is the light it casts on possible sources of these small errors.

The detailed design of the instrument described by Rowe and Abraham has been improved considerably. Davies built a modified version, while Hanson (1961) has described an instrument with electronics sufficiently stable for use with a recorder. The recorder may be calibrated and then used for direct reading of osmotic pressures. Hanson does not recommend the use of controlled back pressure or interpolation, though this method could be used with this type of instrument. A similar instrument has been made commercially available (Meelabs and Knauer osmometers).† A somewhat different approach was used by Rolfson and Coll (1964). Their basic arrangement is shown in Fig. 3.7 and automation has been taken a stage further than in the instruments already described. A servomechanism is used to apply a pressure automatically until the rate of change of pressure as detected by a capacity sensor is zero. This must correspond to zero solvent transport, and hence the applied pressure is the osmotic pressure. An advantage of this apparatus, provided that its response time is fast, is that it minimizes solvent transport, and thus tends to avoid filtration errors. (The Dohrmann-instrument also uses this principle.) Another commercial instrument (Mechrolab) operates on similar principles but instead of a direct electronic pressure-sensing device uses the movement of a bubble held in a capillary. Any movement of the bubble actuates a servomechanism which applies pressures to

† Melabs. (CSM-1 and CSM-2 recording osmometer.) 3300 Hillview Avenue, Stanford Industrial Park, Palo Alto, California U.S.A. (In the U.K. Techmation Ltd., 19 Carlisle Road, London N.W. 9.)
Dohrmann M-150 Stabin/Shell automatic osmometer.
Dohrmann Instruments Co., 990 Varian Street, San Carlos, California 94070, U.S.A. (In the U.K. Techmation Ltd.)
Mechrolab Osmometers.
Several models available. Hewlett-Packard Co., Avondale, Pa. U.S.A. (Branches in various countries)
Knauer membrane osmometer.
H. Knauer and Co. GmbH, 1 Berlin 37, Holstweg 18, Germany. (In the U.K. A.D. Whitehead, Ardleigh, Colchester, Essex.)

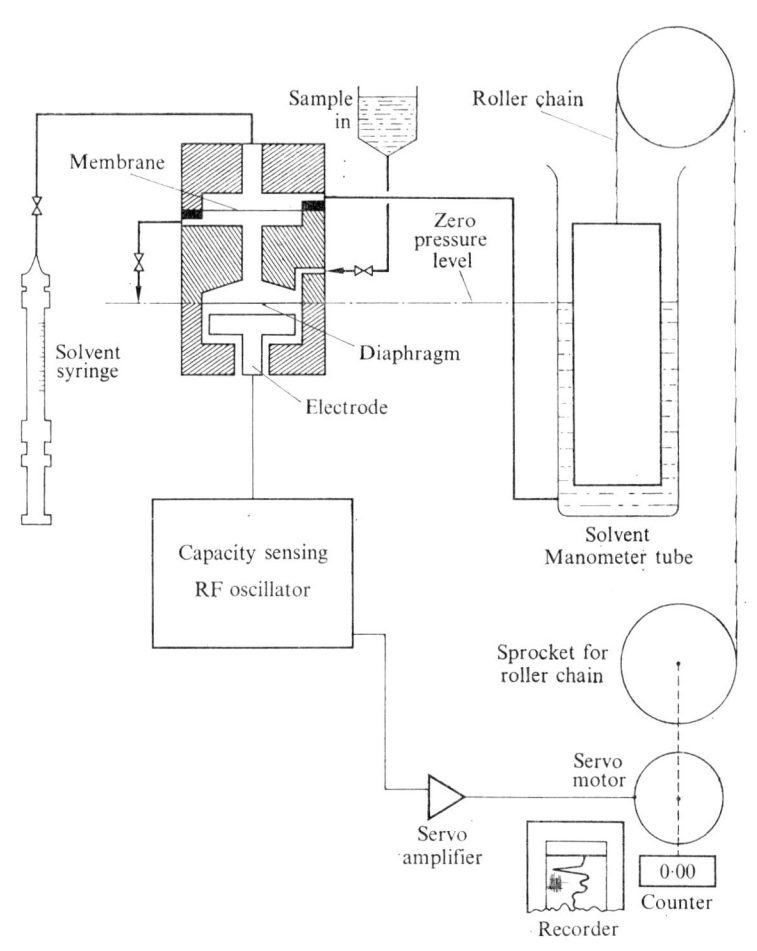

FIG. 3.7. The automatic osmometer of Rolfson and Coll (1964). Deflection of the diaphragm actuates a servomechanism which automatically applies a back pressure until solvent flow through the membrane is zero.

maintain the bubble stationary. Some difficulties have arisen with aqueous solvents because of surface tension effects at the bubble, but this instrument has been successfully used in extended studies of proteins (e.g. Lauffer 1966; Paglini 1968). It is rapid and convenient in use, and is well suited to large numbers of routine estimations. It appears however to be impossible to determine a true buffer zero, which might make small leaks difficult to detect.

3.6. Filtration errors

In the course of experiments designed to measure the constants k_1 and k_2, Davies (1966) discovered yet another source of error which is probably

84   OSMOMETERS AND OSMOTIC PRESSURE MEASUREMENTS

more important than any so far considered. When solvent passes through the membrane, the layers of solution immediately adjacent must be either diluted, or concentrated depending on the direction of solvent flow. These concentration gradients will be eliminated by the usual processes of diffusion and possibly convective flow, but this takes a finite time and until it is

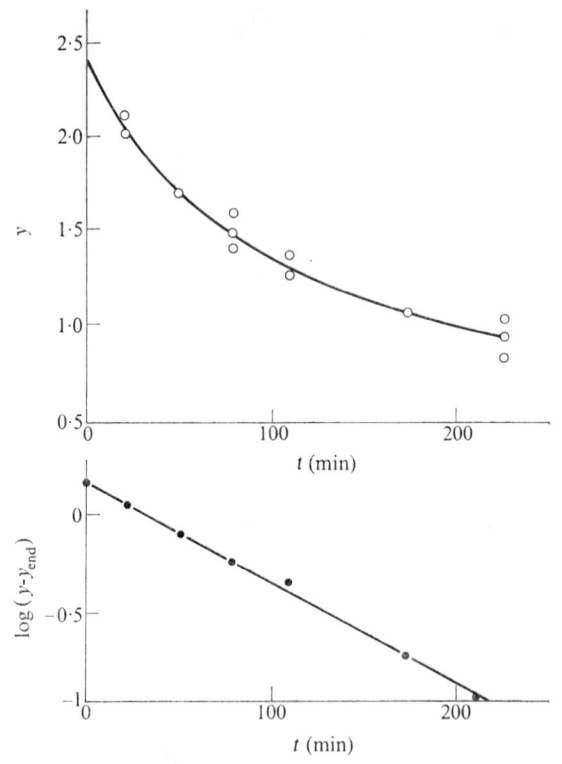

FIG. 3.8. Rate of loss of a filtration error. y is a galvanometer reading, in arbitrary units, where $y = 0$ corresponds to no error. The lower plot shows that it is lost roughly according to first order kinetics. Taken from Davies (1966).

completed the measured equilibrium pressure will not correspond to the true osmotic pressure of the bulk solution. Davies found that the time required for these effects to disappear is longer than the time usually allowed for making measurements. Thus, Fig. 3.8 shows the time course of the disappearance of an error caused by ultrafiltration. The osmometer was first equilibrated, and then, with the tap open, a pressure well above the equilibrium pressure was applied, causing solvent to move from the solution chamber through the membrane. A layer of more concentrated solution must then be in contact with the membrane, though the actual volume transferred would make only a negligible difference to the composition of the bulk solution. A large error appeared, and after 200 minutes had still not disappeared. The rate of disappearance was consistent with a first order relationship.

Filtration in the opposite sense leads to similar errors in the opposite direction. Schultze and Kuhn (1961) and Kuhn (1951) have derived expressions relating the thickness of the layers and the diffusion constants to the rate of re-equilibration. These errors, which may be generally called filtration errors, are undoubtedly widespread, and are responsible for many of the anomalies which have sometimes been attributed to membranes. The only way to avoid them, in osmometers of this type, is to keep solvent transport as low as possible. This is accomplished by estimating the osmotic pressure in advance and applying this pressure as soon as the solution is inserted. Errors of this type may be detected by re-measuring the equilibrium pressure over a period of thirty minutes. Any drift is an indication of filtration errors, on the assumption that temperature drifts and leakages can be confidently eliminated.

Extrapolation could then be used, to allow for such errors, but it is doubtful if all the other sources can be sufficiently controlled to make this worthwhile. Provided that solvent transport is reduced to the minimum needed to make the measurements, and some check is kept on the possibility of filtration error, then no serious problems arise with most solutions. Very viscous samples, because of decreased rates of solute diffusion, may however present problems of a greater magnitude, and instruments where convective mixing or even mechanical agitation is possible may have to be used.

3.7. Osmometers for very low pressures

As has already been pointed out, the simple osmometers described in Section 3.2 are to be preferred when very low pressures have to be measured. (For present purposes very low means less than 5 mm of water, down to 0·001 mm.) There are, however, two special developments which are particularly interesting, both of which replace the manometer–cathetometer method of measuring pressures. The first of these, the osmotic balance, was described by Jullander and Svedburg (1944) and by Enoksson (1948). In this apparatus, the arrangement was as shown in Fig. 3.9. The rate of solvent passage through the membrane was simply determined by weighing the osmometer cell, and the applied pressure was varied by raising or lowering the level O. An ordinary analytical balance could be used, but because of buoyancy effects the cell had always to be weighed at exactly the same level of immersion. Enoksson described an optical arrangement to facilitate this.

There are problems of temperature control, but the most serious drawback is that surface tension effects with aqueous solvents are too great to be corrected. There are also other difficulties, which are connected with evaporation of the solvent and the volume of the cell walls, which with their associated

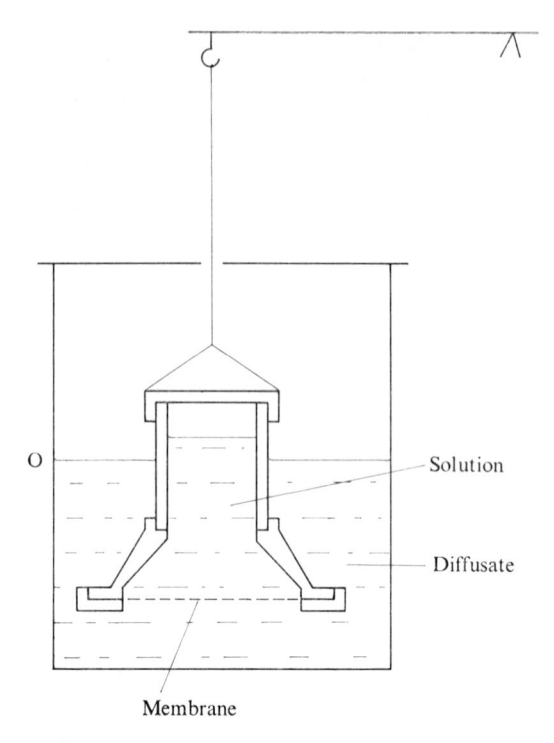

86 OSMOMETERS AND OSMOTIC PRESSURE MEASUREMENTS

FIG. 3.9. The osmotic balance, after Enoksson (1948). The solvent transport is determined by weighing the osmometer.

buoyancy effects reduce the sensitivity. It seems unlikely that this approach would repay further development.

A more promising line is that of Claesson and Jacobsson (1954) who have made an osmometer with a very precise optical method for determining the difference in height of two menisci, which is of course the central problem of measuring low pressures. They used a Fuoss–Mead cell and determine the rate of movement of the meniscus under small applied pressures (from 0·3 to 1·5 cm of water) with an interferometer. The arrangement is shown in Fig. 3.10. The equilibrium pressure was determined in the usual way (Jacobsson 1954) by interpolation and values as low as 0·3 mm of water obtained with apparent ease and precision. Although the solvents used were organic liquids, there appears to be no reason why aqueous solvents should not be used. Jacobsson made use of a 'reference zero', measured with solvent in both parts of the osmometer, and this is standard practice in all osmometry at present. It should be pointed out, however, that at very low pressures, the magnitude of the capillary corrections may be as great as, or even exceed the true osmotic pressure. That is, although equilibrium pressures may be

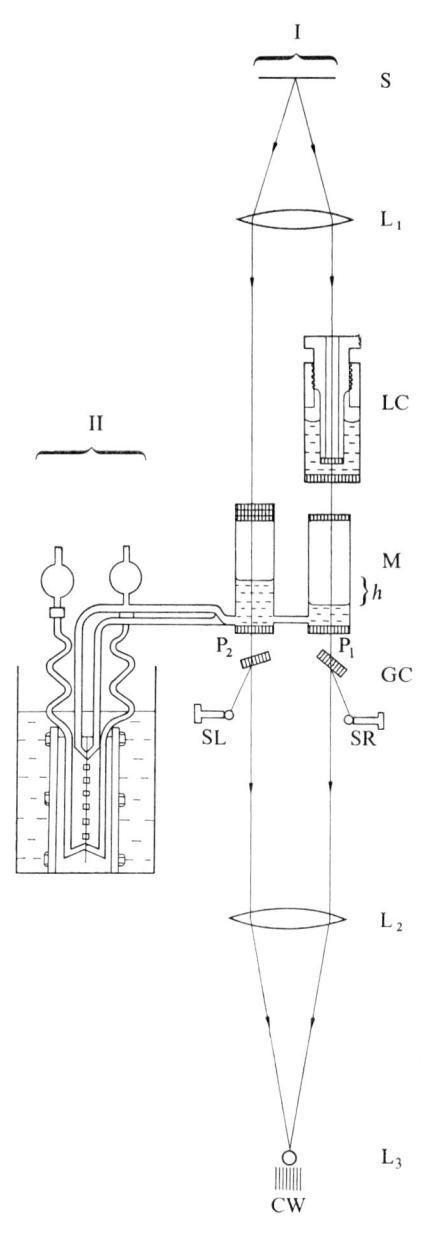

FIG. 3.10. An osmometer for low pressures. (taken from Claesson and Jacobsson 1954). The difference in levels h, in a manometer, are found by interferometric measurements. S, a slit source, L_1 and L_2 convex lenses, L_3 cylindrical lense. LC, P_1 and P_2 compensators.

88   OSMOMETERS AND OSMOTIC PRESSURE MEASUREMENTS

measured, it may not follow directly that these are the 'osmotic pressures' defined in Chapter 2 and thus derived molecular weights may be in error.

3.8. Gel shrinkage methods

Proctor and Wilson in 1916 made measurements of osmotic pressure by observing the shrinkage or swelling of gels. If a gel particle is surrounded by a solution containing solutes which cannot penetrate the gel mesh then the surface layer of the gel becomes effectively a semipermeable membrane, and solvent flows into the particle, causing it to swell. At equilibrium elastic resistance of the gel to distortion is just sufficient to oppose the osmotic pressure. Ogston (1966) discusses the interesting question of how to define the osmotic pressure of a gel: however defined, a possible method of measurement based on dimensional changes of gel particles, will clearly require calibration with solutions of known osmotic pressure. In fact, mammalian cells such as erythrocytes may be used to compare osmotic pressures by observing the concentration of solute needed to bring about rupture, but this method is necessarily confined to a single pressure.

More recently Edmond *et al.* (1968), Ogston and Wells (1970) and Ogston and Preston (1973) have re-examined the possibility of using calibrated gel particles as osmometers. Their results suggest that these methods ought to be considered when choosing a technique for osmotic pressure measurement. Ogston's first method employs beads of 'Sephadex' a commercially-available cross-linked dextran gel. Sephadex G50 is not penetrated by proteins with molecular weight of greater than 100 000, which sets an effective lower limit for the size of solute.

The bead is impaled on a needle and the change in its diameter from an initial value to the measured value is observed by microscopy, often with the aid of photomicrographs. The shrinkage caused by immersion in solution appeared to be completely reversible on reversion to solvent so that a single calibrated bead could be used many times—a distinct practical advantage. Fig. 3.11 shows the time course of shrinkage in dextran and polyethylene glycol solutions, and equilibration is clearly rapid and complete in about ten minutes. PEG 6000 penetrates the bead, and the deviations from a smooth curve are due to the different rates of penetration of PEG and water. Note that penetration does not prevent equilibrium shrinkage from being obtained, though of course the corresponding pressure will be dependent on the degree of penetration.

Fig. 3.12 shows a non-linear calibration curve, suggesting a precision of about 1·5 cm H_2O. The method is most effective for pressures of the order of 1000 cm H_2O. It suffers from the drawbacks that the gel is penetrated by many macromolecules of interest, and the difficulty of determining the volume of small beads from measurements of their diameter. These problems have been solved in a new development, employing polyacrylamide gels and a

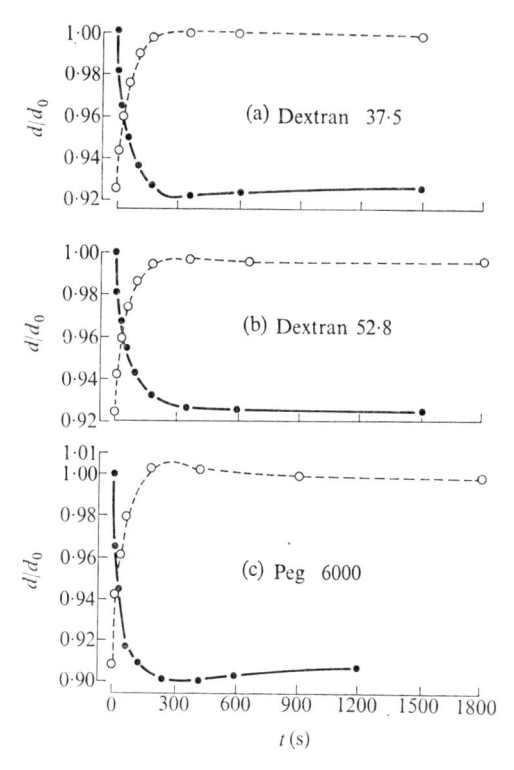

FIG. 3.11. Bead diameter of a bead of Sephadex G-50, where d is the diameter and d_0 the initial diameter plotted against time of immersion in solution (●) and then in water (○). (a) Dextran 37·5, 0·094 g/ml. (b) Dextran 52·8, 0·091 g/ml. (c) Polyethyleneglycol 6000, 0·048 g/ml.

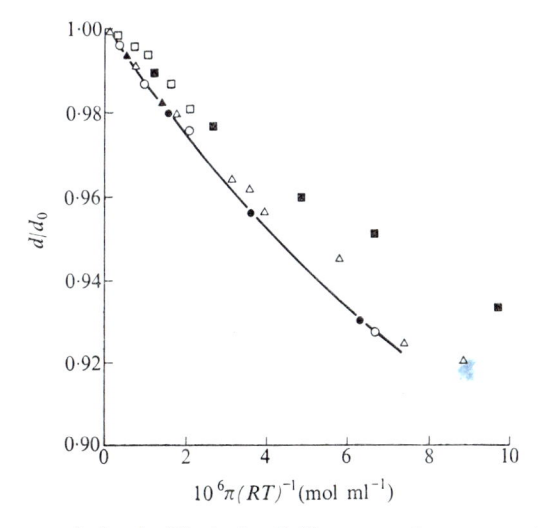

FIG. 3.12. Diameters of a bead of Sephadex G-50, expressed as a proportion of the initial diameter, as a function of the osmotic pressure of the solution. ● Polyethyleneglycol 20 000, ○ Dextran 500, ▲ Dextran 52·8, △ Dextran 37·5, ■ Dextran 19·7, □ Bovine serum albumin, all in 0·2 M NaCl.

strip of gel. (Ogston and Preston 1973.) The acrylamide gel is cast onto strips of cellulose tissue in the form of a loop. The loop behaves when shrinking much as a bimetallic strip responds to temperature changes, and by measuring the distance between the ends of the loop a pressure sensitivity comparable to that of good membrane osmometers can be obtained. Also, acrylamide gels can be made with concentrations up to 70 per cent, and are capable of excluding solutes with molecular weights of about 10 000 or less. Excellent values for the molecular weights of cytochrome C (12 500), ovalbumin (44 000) and human serum albumin (66 000) were obtained with this method and it is clearly able to yield results comparable with those from membrane osmometers. There remains the practical problem that the loop must be calibrated. Although many precise molecular weights for proteins are known, less published data are available for the virial coefficient.

3.9. Membranes

It will be evident from the preceding discussion that modern osmometers require rather more from a membrane than merely semi-permeability. For the membrane must be strong, inelastic, reasonably uniform and above all it must possess adequate permeability to the solvent.

In almost all the earlier work collodion (cellulose nitrate) membranes were used. Philipp and Djork (1951) made a systematic study with different cellulose acetate and collodion membranes in contact with a cellulose acetate solution. Some at least of the cellulose acetate was capable of passing through the membranes, but they concluded that there was no systematic connection between the permeability to solvent and to cellulose acetate. They also attempted to measure pore diameters in the membranes and found, as they expected, that the solvent permeability was related to the pore size but that the osmotic pressure developed was not. Patat (1959) has made an extensive study of membranes useful in both aqueous and organic solvents and his article should be consulted for a survey of available materials and estimates of their permeabilities.

Vaughan (1959) describes methods for making cellulose acetate membranes, said to be at least the equal of commercially available types, by casting solutions of cellulose acetate containing n-amyl alcohol (n-pentan-1-ol) as a precipitant. He also (1958a, b) describes techniques for modifying the permeability and although these were mainly intended for work in organic solvents it is probable that such treatments would also affect the water permeability.

Craig *et al.* (1957) give a useful description of methods of modifying the permeability of cellophane membranes, while Renkin (1954) gives a thorough account of permeability by relatively small molecules.

Perhaps the most exotic method of preparing membranes was that described by Masson and Melville (1949), who grew a bacterium (*Acetobacter Xylinum*) which, in suitably shaped vessels, deposits a disc of cellulose. They used it successfully as a membrane in their Fuoss–Mead osmometers.

Some very complicated methods for preparing collodion membranes have been published. A simple and effective one is to dissolve about 1–2 per cent cellulose nitrate in a 50:50 mixture of ether and ethanol. (Commercial samples vary—a suitable one is I.C.I. Type HL 120-170.) Castings of thin films are then prepared by dipping a clean boiling-tube, or sheet of glass into the solution, withdrawing into the solvent vapour, allowing to drain and then drying completely. This does require a little practice to obtain uniform films. After they are dry they can be peeled off the glass: alternatively for more complicated shapes Smithies (1953) method of casting onto sodium thiosulphate formers, which can subsequently be dissolved away in water can be used.

Before use the membranes are re-swollen in alcohol–water mixtures in the range 95–98 per cent ethanol. The higher the alcohol content the lower the water permeability. Membranes can be kept indefinitely in these mixtures, equilibration is rapid and they rapidly equilibrate with the required buffer on immersion in it. Trial and error is required to find a membrane with a suitable permeability: as already pointed out this is routinely measured as a part of osmotic pressure determination so a separate characterization by permeability measurements is not required. This method is adapted from Wells (1932) who describes a more elaborate technique.

Most workers will probably wish to avoid the complications of preparing membranes, and will at least attempt to use commercially available types. There seem to be two major suppliers at present.† From the characteristics claimed by the manufacturers Amicon PM10 would probably be the most suitable (the composition of the membranes is not disclosed). For molecular weights greater than 30 000 PM30 might be better, since it has a higher water permeability. Both manufacturers make a wide range of membranes of differing pore sizes, primarily designed for pressure filtration apparatus. Sartorius recommend their type 'UCF very dense' and 'UCF super dense', made from regenerated cellulose. They also manufacture cellulose nitrate and cellulose acetate based membranes.

Looking to the future it is possible that supported glass membranes may become available, though these would almost certainly have adsorption problems with proteins. It may also be possible to make membranes from suitably supported polyacrylamide gels. Membranes are available bearing

---

† Amicon N.V., Meckelaarstraat, Oosterhout (N.B.), Holland. (In the U.K., 57 Queens Road, High Wycombe, Bucks.)
Sartorius Membranfilter GmbH, 34 Goettingen, Weender Landstrasse 96–102, West Germany. (Agents in several countries: in the USA, Carl Schleicher & Schuel Co., Keene, New Hampshire; in the UK, V. A. Howe & Co., Pembridge Road, London W11.)

ionic groups, but these would introduce complications into work with polyelectrolytes. Providing a membrane is semi-permeable, it does not matter in principle whether it bears charges or not, but there is a danger that charged solutes would interact with the membrane itself, perhaps changing its characteristics. It is possible however that suitable manipulation of pH could result in a very low permeability to polyelectrolytes and a high permeability to water.

There can be little doubt that good osmometry depends on a good membrane and much emphasis has been placed on this in the past, though the internal checks required by more recent instruments ensure that it is almost impossible to obtain measurements at all unless the membrane is adequate. It is arguable that the membrane is now the part of the osmometer most in need of further development, and the appearance of reliable, reproducible membranes would do much to make osmometry a more attractive option in work on macromolecular solutions.

3.10. Typical calculations from osmotic pressure observations

Molecular weight and virial coefficients

Sometimes points of difficulty arise in handling results which are not adequately dealt with by a more general discussion. This section describes, therefore, a typical set of results and details of methods of obtaining the molecular weight, the first virial coefficient and an estimate of the precision.

Osmotic pressure measurements were made on the protein glycinin, a major globulin of the soybean. The native protein has a molecular weight of 363 000 by sedimentation–diffusion methods, while light scattering studies indicated a value of 345 000. The protein is known to have three different N-terminal amino-acids (glycine, phenylalanine, and valine) and the ultimate objective was to determine the number of sub-units in the protein. For reasons already given (Section 2.6) we require the number average molecular weight of the fully dissociated protein.

The following results were obtained (Tombs, unpublished observations) with glycinin dissolved in 4 M guanidinium hydrochloride, 0·1 M sodium sulphite solution at 25°C. The first step is to plot these results in the form shown in Fig. 3.13. Inspection shows reasonable linearity, so that a single virial coefficient (B or its equivalent, see Table 1.1) is probably adequate to describe the line and we can therefore use eqn (2.4) in the form with higher coefficients taken as zero. So

$$\frac{\pi}{c} = RT\left[\frac{1}{M_2} + Bc_2\right] \qquad (2.4a)$$

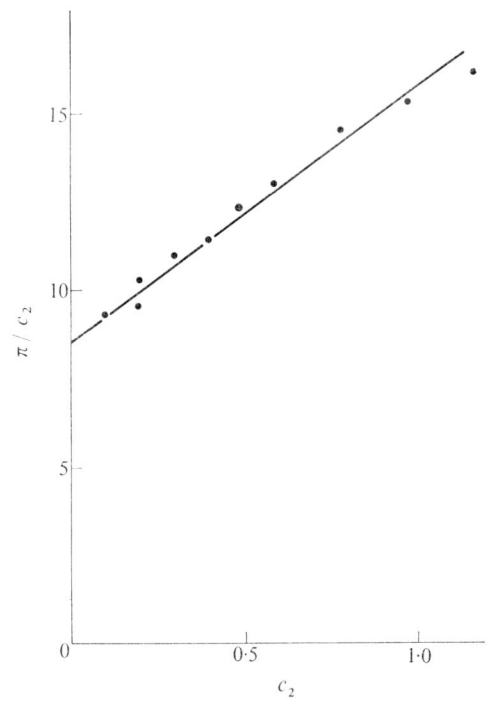

FIG. 3.13. A typical π/c_2 versus c_2 plot, for glycinin in guanidine hydrochloride solution. See Section 3.10.

TABLE 3.1
Osmotic pressure of glycinin in guanidinium hydrochloride–sulphite solutions

c_2 (concentration g/100 ml solution)	π(cm H$_2$O).	π/c_2	$(c_2)^2$
0·097	0·902	9·3	0·009409
0·195	2·00	10·3	0·038025
0·195	1·85	9·5	0·038025
0·292	3·21	11·0	0·085264
0·390	4·45	11·4	0·1521
0·487	5·99	12·3	0·237169
0·585	7·60	13·0	0·342225
0·780	10·14	14·5	0·6084
0·975	14·91	15·3	0·950625
1·170	18·83	16·1	1·3689
Total 5·166		122·7	3·830142

94 OSMOMETERS AND OSMOTIC PRESSURE MEASUREMENTS

or, in this case we choose

$$\frac{\pi}{c_2} = \frac{RT}{M_2}[1+B'c_2] \qquad (2.4b)$$

noting that

$$B' = BM_2$$

One treatment of the results though not the best, is as follows: Putting

$$\frac{\pi}{c_2} = y \qquad c_2 = x \qquad a = \frac{RT}{M_2} \qquad b = \frac{RTB'}{M_2}$$

then

$$y = a+bx \qquad (3.20)$$

The residual standard deviation of y on x is

$$s^2 = \frac{1}{n-2}\left\{\left(\sum(y-\bar{y})^2 - \frac{[\sum(x-\bar{x})(y-\bar{y})]^2}{\sum(x-\bar{x})^2}\right)\right\} \qquad (3.21)$$

where n is the number of determinations, and a bar here and elsewhere indicates an arithmetic mean.

$$b = \frac{\sum(x-\bar{x})(y-\bar{y})}{\sum(x-\bar{x})^2} \qquad (3.22)$$

while

$$a = \bar{y} - b\bar{x}$$

and

$$B' = \frac{b}{a} \qquad (3.23)$$

It is convenient to use

$$\sum(x-\bar{x})^2 = \sum x^2 - \frac{(\sum x)^2}{n} \qquad (3.24)$$

$$\sum x^2 = 3\cdot 830 \qquad \frac{(\sum x)^2}{n} = 2\cdot 668$$

so

$$\sum(x-\bar{x})^2 = 1\cdot 162$$

Similarly

$$\sum y^2 = 1557\cdot 63 \qquad \frac{(\sum y)^2}{n} = 1505\cdot 50$$

so

$$\sum(y-\bar{y})^2 = 52\cdot 0$$

also

$$\sum xy - \frac{\sum x \sum y}{n} = \sum(x-\bar{x})(y-\bar{y}) \qquad (3.25)$$

and

$$\sum xy = 71\cdot 08 \qquad \frac{\sum x \sum y}{n} = 63\cdot 386$$

$$\sum(x-\bar{x})(y-\bar{y}) = 7\cdot 694$$

OSMOMETERS AND OSMOTIC PRESSURE MEASUREMENTS 95

$$\therefore b = \frac{7 \cdot 694}{1 \cdot 162} = 6 \cdot 62$$

$$a = 12 \cdot 27 - 6 \cdot 62 \times 0 \cdot 5166$$
$$= 8 \cdot 85$$

$$B' = \frac{6 \cdot 62}{8 \cdot 85} = 0 \cdot 74$$

$$\pi/c_2 = 8 \cdot 85 + 6 \cdot 62 c_2 \tag{3.26}$$

Now to calculate s,

$$s^2 = \frac{1}{8}\left(52 \cdot 10 - \frac{(7 \cdot 69)^2}{1 \cdot 1613}\right)$$
$$= 0 \cdot 121$$

and
$$s = 0 \cdot 349$$

and
$$a = 8 \cdot 85 \pm 0 \cdot 349$$

The value of 8·85 is slightly higher than that obtained by the graphical extrapolation shown in Fig. 3.13 which is about 8·6. A statistical procedure of the type outlined is often used but is not the best available. As c decreases, the precision of π probably decreases so that the error of π/c shows a systematic change with c, and the points should be weighted by c. It is thus preferable to use the rather more complicated fit of

$$\pi = \alpha + \beta c + \gamma c^2 \tag{3.27}$$

where α is the potentially non-zero value for π when $c = 0$. Putting $\pi = y$, $c = x_1$ and $c^2 = x_2$, one then obtains the best fit of eqn (3.28) by Fisher's c-matrix method.

$$(y - \bar{y}) = b(x_1 - \bar{x}_1) + b_2(x_2 - \bar{x}_2). \tag{3.28}$$

The total number (n) of observations now includes one for $\pi = 0$, $c = 0$, and so $n = 11$, for the data of Table 3.1. The quantities b_1 and b_2 are estimated from the use of the following expressions in sequence and in combination.

$$D = (\sum (x_1^2) - n(\bar{x}_1)^2)(\sum (x_2^2) - n(\bar{x}_2)^2) - (\sum (x_1 x_2) - n\bar{x}_1 \bar{x}_2)^2 \tag{3.29}$$

$$C_{11} = \{\sum x_2^2 - n(\bar{x}_2)^2\}/D \tag{3.30a}$$

$$C_{12} = \{\sum (x_1 x_2) - n(\bar{x}_1 \bar{x}_2)\}/D \tag{3.30b}$$

$$C_{22} = \{\sum (x_1)^2 - n(\bar{x}_1)^2\}/D \tag{3.30c}$$

$$b_1 = (\sum (x_1 y) - n\bar{x}_1 \bar{y})C_{11} + (\sum (x_2 y) - n\bar{x}_2 \bar{y})C_{12} \tag{3.31}$$

$$b_2 = (\sum (x_1 y) - n\bar{x}_1 \bar{y})C_{12} + (\sum (x_2 y) - n\bar{x}_2 \bar{y})C_{22} \tag{3.32}$$

The residual variance R is given by

$$R = (\sum (y^2) - n(\bar{y})^2) - (\sum (x_1 y) - n(\bar{x}_1 \bar{y}))b_1 - (\sum (x_2 y) - n(\bar{x} \bar{y}))b_2. \tag{3.33}$$

96 OSMOMETERS AND OSMOTIC PRESSURE MEASUREMENTS

While the standard errors (s.e.) of the estimate of b_1 and b_2 are

$$\text{s.e. of } b_1 = \left(\frac{Rc_{11}}{n-3}\right)^{\frac{1}{2}} \quad (3.34)$$

$$\text{s.e. of } b_2 = \left(\frac{Rc_{22}}{n-3}\right)^{\frac{1}{2}}. \quad (3.35)$$

The quantity α is given by

$$\alpha = \bar{y} - b_1\bar{x}_1 - b_2\bar{x}_2 \quad (3.36)$$

and its standard error is given by

$$s^2 = \frac{R}{n-3}\{1+(\bar{x}_1)^2 C_{11}+(\bar{x}_2)^2 C_{22}\} \quad (3.37)$$

The quantities calculated from the data in Table 3.1 are as follows:

$\sum y = 69{\cdot}882$ $\qquad \sum x_1 = 5{\cdot}166 \qquad \sum x_2 = 3{\cdot}830142$

$\bar{y} = 6{\cdot}352909 \qquad \bar{x}_1 = 0{\cdot}49636 \qquad \bar{x}_2 = 0{\cdot}348195$

$n(\bar{y})^2 = 443{\cdot}95398 \qquad n(\bar{x}_1)^2 = 2{\cdot}426140 \qquad n(\bar{x}_2)^2 = 1{\cdot}333635$

$\sum (y^2) = 811{\cdot}6794 \qquad \sum (x_1^2) = 3{\cdot}830142 \qquad \sum (x_2)^2 = 3{\cdot}354478$

$\sum (y^2) - n(\bar{y})^2 \qquad \sum (x_1^2) - n(\bar{x}_1)^2 \qquad \sum (x_2)^2 - n(\bar{x}_2)^2$

$= 367{\cdot}7254 \qquad = 1{\cdot}404002 \qquad = 2{\cdot}020843$

$\sum (x_1 x_2) = 3{\cdot}418686 \qquad n(\bar{x}_1 \bar{x}_2) = 1{\cdot}798775$

$\sum (x_1 x_2) - n(\bar{x}_1 \bar{x}_2) = 1{\cdot}619911$

$\sum (x_1 y) = 55{\cdot}351344 \qquad \sum (x_2 y) = 51{\cdot}246359$

$n(\bar{x}_1 \bar{y}) = 32{\cdot}819128 \qquad n(\bar{x}_2 \bar{y}) = 24{\cdot}332544$

$\sum (x_1 y) - n(\bar{x}\bar{y}) = 22{\cdot}532616$

$\sum (x_2 y) - n(\bar{x}_2 \bar{y}) = 26{\cdot}913815$

From eqn (3.29), $D = 0{\cdot}213156$, so

$$C_{11} = \frac{2{\cdot}020843}{0{\cdot}213156} = 9{\cdot}480584$$

$$C_{12} = \frac{1{\cdot}619911}{0{\cdot}213156} = 7{\cdot}59965$$

$$C_{22} = \frac{1{\cdot}404002}{0{\cdot}213156} = 6{\cdot}58673$$

Then, from eqn (3.31)

$$b_1 = (22{\cdot}532616 \times 9{\cdot}480584) - (26{\cdot}913815 \times 7{\cdot}59965)$$
$$= 213{\cdot}62235 - 204{\cdot}535574$$
$$= 9{\cdot}086758.$$

Similarly, from eqn (3.32)
$$b_2 = 6{\cdot}034176$$

From eqn (3.33) $R = 0{\cdot}574275$ and from this and eqns (3.34) and (3.35) the standard errors of b_1 and b_2 are

$$b_1 = 9{\cdot}0867 \pm 0{\cdot}8249 \qquad b_2 = 6{\cdot}03417 \pm 0{\cdot}6876$$

α becomes, from eqn (3.36),

$$6{\cdot}352909 - 9{\cdot}086758 \times 0{\cdot}4696 - 6{\cdot}03417 \times 0{\cdot}34819 = -0{\cdot}0156$$

and its standard error is 0·528 so,

$$\alpha = -0{\cdot}0156 \pm 0{\cdot}528$$

Thus,
$$\pi = -0{\cdot}0156(\pm 0{\cdot}528) + 9{\cdot}086(\pm 0{\cdot}825)c + 6{\cdot}034(\pm 0{\cdot}688)c^2$$

From
$$\pi = \frac{RT}{M}c + \frac{RTB'}{M}c^2, \qquad (2.4b)$$

One obtains
$$RT/M = 9{\cdot}086 \pm 0{\cdot}825 \text{ cm H}_2\text{O g}^{-1} \text{ dl}$$

and
$$B' = \frac{6{\cdot}034}{9{\cdot}086} = 0{\cdot}664(\pm 0{\cdot}15) \text{ dl g}^{-1}$$

It should be noted that the intercept at $c = 0$ does not differ significantly from $\pi = 0$. These values, although slightly different from those obtained by the simple treatment will give a better estimate of the best value, and the error. We now require a value for RT, which is usually quoted in units such as
$$R = 0{\cdot}08204 \text{ l atm mol}^{-1} \text{ deg}^{-1}$$
$$= 8{\cdot}314 \times 10^7 \text{ erg mol}^{-1} \text{ deg}^{-1}$$
$$= 8{\cdot}314 \text{ J mol}^{-1} \text{ deg}^{-1}$$
$$= 1{\cdot}987 \text{ cal mol}^{-1} \text{ deg}^{-1}$$

and clearly the units chosen affect the numerical value. All these values are re-statements of the fundamental observation that 1 mole of a substance, dissolved in 1 dl (= 100 ml) gives an osmotic pressure of 224·1 atmosphere at 273 K. At 25°C it gives a pressure of $224{\cdot}1 \times \frac{298}{273}$ atmosphere, and since

98 OSMOMETERS AND OSMOTIC PRESSURE MEASUREMENTS

one atmosphere is the pressure exerted by 1033·2 cm water, the required value is

$$RT = 224 \cdot 1 \times \tfrac{298}{273} \times 1033 \cdot 2 = 2 \cdot 527 \times 10^5 \text{ dl cm } H_2O \text{ mol}^{-1}$$

with concentrations in g dl^{-1} of solution, and pressures in cm of water.
Hence

$$M_2 = \frac{2 \cdot 527 \times 10^5}{9 \cdot 086} = 27\,812 \text{ g mol}^{-1},$$

while 9·086±0·825 yields values of 30 589 and 25 496. The final result is therefore

$$M_2 = 27\,812 \pm 2800 \text{ g mole}^{-1}$$

We also have a value for B', and remembering that it is equivalent to BM_2 in the nomenclature of chapter 2 (units are discussed below):

$$B' = BM_2 = 0 \cdot 66 \text{ dl g}^{-1}$$

and so

$$B = 2 \cdot 37 \times 10^{-5} \text{ mole dl g}^{-2}$$

The main interest in the value of B is to compare it with other values obtained in different solvents or with different proteins and it is necessary to be quite sure that all the values are expressed in comparable units.

The relevant term in eqn (2.4b) for B' is $(RT/M_2)B'c_2$ from which it is clear that $B'c_2$ is a pure number, so that the units of B' are those of reciprocal concentration, in this case, dl g^{-1}. Since B is B'/M_2, the units of B are mole dl g^{-2} since

$$\frac{B'}{M} = \frac{\text{dl g}^{-1}}{\text{g mol}^{-1}} = \text{mol dl g}^{-2}$$

Although B' and B are often quoted in decilitres, it is becoming customary to express virial coefficients in terms of concentrations/ml so that B' becomes ml g^{-1} and B mol ml g^{-2}.

Thus in this case $B' = 0 \cdot 66$ dl g$^{-1} = 66$ ml g^{-1} and B becomes $2 \cdot 37 \times 10^{-3}$ mol ml g^{-2}. Although the 'practical' units of g dl^{-1} are frequently used, we can now see that units would have been simpler if concentrations had been expressed in g ml^{-1} from the start.

The result and the effect of converting the concentration scales can easily be checked. Thus, by direct measurement on the graph in Fig. 3.13, and altering the concentration scale to g ml^{-1}, the slope of the line of π/c_2 vs c_2 is $6 \cdot 4 \times 10^2/0 \cdot 9 \times 10^{-2} = 7 \cdot 11 \times 10^4$ cm H$_2$O ml g^{-2}. The value of RT must also be adjusted to allow for the new concentration scale, and becomes $2 \cdot 527 \times 10^7$ ml cm H$_2$O mole^{-1}. Therefore

$$\frac{RT}{M_2} = \frac{2 \cdot 527 \times 10^7}{2 \cdot 7812 \times 10^4} = 0 \cdot 908 \times 10^3 \text{ cm H}_2\text{O g}^{-1} \text{ ml}$$

and

$$B' = \frac{7 \cdot 11 \times 10^4}{0 \cdot 908 \times 10^3} = 78 \text{ ml g}^{-1}$$

which is reasonably close to the value of 66 ml g^{-1} obtained above. A value of B of $2\cdot37\times10^{-3}$ mole ml g^{-2} is of the order expected for proteins in a relatively random configuration, and this would be expected for the solvent used. A much smaller value, of the order of 10^{-4} would be appropriate for a compact globular structure. B' is often interpreted as an excluded volume (see Section 2.7) so that its units of volume per mass are appropriate; it is of interest that the value found, 66 ml g^{-1}, is also larger than would be expected for a compact globular structure.

Derivation of number of sub-units

Values for M_2, the number average molecular weight and of B' for glycinin in a variety of dissociating solvents are shown in Table 3.2. They were

TABLE 3.2

Number average molecular weights of glycinin in dissociating solvents

Solvent	M_2	B'(ml g^{-1})	n
2 M urea 0·1 M SO$_3^{2-}$	106 100±1400	15	3·4
4 M urea 0·1 M SO$_3^{2-}$	58 550±4000	30	6·2
6 M urea 0·1 M SO$_3^{2-}$	29 230±1000	17	12·4
8 M urea 0·1 M SO$_3^{2-}$	30 360±500	59	12·0
4 M GHCl 0·1 M SO$_3^{2-}$	27 812±2800	66	14·0
6 M GHCl 0·1 M SO$_3^{2-}$	28 180±900	72	12·9
8 M urea 0·2 M SO$_3^{2-}$	30 030±600	18	12·1
8 M urea	144 000±2000	44	2·5

all obtained by osmometry, while n, the number of units produced was calculated from the 'intact' molecular weight of 363 000, by dividing this value by the determined M_2. Because M_2 is a number average molecular weight, in for example, the case considered in detail above, this gives a value of 14·0 sub-units and this is so whether the sub-units are identical or all different. (A weight average would not give this result.) Leaving aside any uncertainties in the molecular weight of the intact molecule it is fair to ask what the effect of errors in the determined M_2 will be.

The calculated residual standard deviation is, unfortunately, not a complete estimate of the absolute precision of the value. A statistical analysis of the type used is a test of the validity of using only one virial coefficient: marked deviation of the π/c_2 against c_2 plot from linearity would have given relatively large residual standard deviations. The latter is also a good measure of the internal consistency of the results but the largest single source of error is almost certainly in the absolute determination of protein concentrations.

All the solutions used were prepared by dilution of a standard and any errors in the value for this concentration and the dilution factors are therefore translated directly into errors in the molecular weight.

Protein determinations are not easy and ultimately depend on gravimetric measurement of dried protein. They are unlikely to be accurate to better than ± 3 per cent. An estimate of error therefore requires that this should be added to the errors already calculated. (Sedimentation–equilibrium measurement for the determination of molecular weight requires a knowledge of $(1/c)(dc/dr)$, and this does not need an absolute concentration determination. It might be thought that this gives an advantage in precision, but this method also requires a knowledge of $\bar{v}$, the partial specific volume, which does require concentration determination for its measurement. Osmotic pressure measurements do not require $\bar{v}$ explicitly but it is in fact a contribution to the value of B', so the two methods do not differ in this respect or in the need for precise concentration determination.)

Continuing with the same example, we thus obtain extreme values for M_2 of 25 496–30 589 and sub-unit values for n of 14·2–11·9. Now taking account of the uncertainty in the intact molecular weight, we had two values, 345 000 and 363 000. Taking the former, n becomes 13·5–11·3, so the probable maximum range for n is 14·2–11·3 and the uncertainty is about $\pm 1 \cdot 5$ sub-unit.

Taking all the results together, since n must be integral for a fully dissociated protein we may reasonably conclude that the intact molecule contains twelve sub-units. In the case of glycinin there is some ancillary evidence that the intact molecule can dissociate reversibly to two equal sized fragments, which means that n is probably also an even number, so that although $n = 12 \pm 1 \cdot 5$ from the direct measurements, it probably does not contain either eleven or thirteen sub-units.

Strictly, since we know only the number of sub-units and not how many sorts there are (though there must be at least three sorts from the N-terminal amino acids) it is possible that the molecule could dissociate to give two fragments of equal molecular weight but containing different numbers of sub-units. This possibility seems remote against the general background of sub-unit structures so far established. With available precision it is unlikely that direct determination of molecular weights, without ancillary evidence, could give a decisive result when n, the number of sub-units, is greater than twelve. The problem is of course less acute when the number of sub-units is only four or six. This limitation is not confined to osmotic pressure measurements; it applies to any method (e.g. light scattering) which requires absolute determination of protein concentration.

It might be argued that since n is a ratio then errors from this source in the intact molecular weight and in the dissociated protein would be eliminated. A carefully designed experiment might reduce them but usually a variety of solvents must be used, one of them being a dissociating solvent, and concentration determination would still be a major source of error. Concentration errors have a larger effect on B' than on M_2, and although quite elaborate calculations are possible in principle (e.g. of the axial ratio of the molecule) B' is not usually sufficiently precise to support them. It gives a broad indication of the configuration. It should be remembered that Donnan effects are also included in the measured B'. These are minimized in most procedures but must be taken into account in detailed interpretation.

Interchain link frequencies

One further type of calculation is possible. In 8 M urea n for glycinin was found to be 2·52 and comparison with its value in 8 M urea 0·1 M sulphite, *viz* 11·95 suggests that the sub-units were linked together by disulphide interchain links. In a partly dissociated protein of course n need not be integral. In a solvent of 2 M urea and 0·1 M sulphite, for example, n for glycinin was 3·4; in this case the interchain links were probably still those present in the intact protein and probably not disulphides.

Taking the molecular weight of the sub-units as 30 000 the number average degree of aggregation of the sub-units in 8 M urea is

$$(144 \times 10^3)/(3 \times 10^4) = 4·8.$$

That is, each particle contains on average 4·8 chains. The corresponding value in 2 M urea 0·1 M sulphite is 3·18. In 8 M urea therefore there was at least $4·8-1 = 3·8$ interchain links and $3·8/4·8 = 0·79$ interchain links per chain. These calculations illustrate the use of the values of molecular weights in aggregation studies: the average interchain link frequency is useful in measuring the formation of strands during approach to gelation. A fully connected gel must have at least 1·0 interchain links per chain.

3.11. Units and some terms

Units of pressure

The fundamental unit of pressure in the SI system is the Pascal, (Pa), defined as 1 Newton per square metre (1 N m^{-2}). This is equivalent to 10 dyne cm^{-2}. The following conversion units are useful, since most measurements are in practice made in manometers containing water or mercury

$$1 \text{ g cm}^{-2} = 9·678 \times 10^{-4} \text{ atmosphere}$$
$$= 0·735 \text{ mm mercury at } 0°C$$
$$1 \text{ cm of water at } 4° = 980·64 \text{ dyne cm}^{-2}$$
$$1 \text{ dyne cm}^{-2} = 0·1019 \times 10^{-2} \text{ cm water at } 4°$$
$$1 \text{ Pa} = 1·019 \text{ mm water at } 4°$$

1 atmosphere = pressure exerted by 760 mm of Hg at a density of

$$13\cdot595 \text{ g cm}^{-3} = 1\cdot01325 \times 10^6 \text{ dyne cm}^{-2}$$
$$= 1033.2 \text{ cm water at } 4°.$$

The value of R

The value of R must be calculated to be appropriate to the pressure and concentration units employed. Thus if π is in Pa, and the concentration is in g mole cm^{-3} $R = 8\cdot3144 \times 10^6$ Pa cm^{-3} deg^{-1} mole^{-1}. A convenient method is to use the original determination that 1 mole dissolved in 22·41 litre of solution gives a pressure of one atmosphere. Thus, with concentrations in g dl^{-1} solution and pressures in cm water, at 25°C, $RT = 2\cdot527 \times 10^5$ dl cm H$_2$O mole^{-1}. This will usually be the most useful value in practice.

Terms

A number of terms are in use, particularly in physiological work, that have not been employed here. They often arise in connection with other colligative properties of solutions such as vapour pressure or freezing point depression and indeed instruments for the measurement of such properties are sometimes described as 'osmometers'. In the sense that the results obtained from them can be expressed in osmotic pressure terms, as can any colligative property, this is perhaps justified, though no-one has proposed to call membrane osmometers, 'vapour-pressure machines' which could be equally justified by the same argument.

Osmolality and osmolarity

This is particularly common in work on electrolytes (Wolf 1966). On either the molal, or molar scales the osmolality is the molar (or molal) concentration which would give the observed osmotic pressure, or other colligative property, if present in ideal solution. Put another way, the osmotic pressure is used to calculate the molality of the solution assuming the simple Van't Hoff equation to apply.

In a typical example, 1 molal sodium chloride solutions give osmotic pressures (actually measured by freezing point depression) corresponding to 1·86 molal ionic concentration rather than the ideal value of 2·0. The ratio between the observed osmolality, and the ideal is the osmotic coefficient, which is defined in Chapter 2. A difficulty with osmolality is that its value depends on the value of the osmotic coefficient, which is another form of activity coefficient and is not of course a constant. The use of 'osmolarity' and 'osmolality' is really an attempt to determine the total concentration of osmotically active substances in physiological fluids and is bound to have limited precision, because the van't Hoff equation is used.

Osmosity

In an attempt to avoid this problem, the 'osmosity' is sometimes used. This is merely the concentration of a sodium chloride solution which gives an identical osmotic pressure, again always measured in practice by freezing point depression.

'Crystalloid pressure', 'Colloid osmotic pressure'

These terms are now little used, but again derive from physiological situations where it was desired to distinguish between the osmotic pressure due to small electrolytes (crystalloids) and that due to macromolecules—the 'colloid' osmotic pressure. Since there is no fundamental distinction to be made it is unlikely that these terms will be revived.

Isotonic, hypertonic, hypotonic, and tonicity

Mammalian cells swell or shrink when placed in solutions, and the extent of shrinkage or swelling is related to the difference of osmotic pressure between the solution and the contents of the cell. The cell membrane behaves as a semi-permeable membrane. Thus, the terms hyperosmotic, hypo-osmotic and iso-osmotic are sometimes used. The interpretation of such quantities is difficult because the cell membrane is rarely perfect (see Section 2.9) and the nature of the solutes will also affect the results (Olmstead 1966).

4

APPLICATIONS OF OSMOTIC PRESSURE MEASUREMENTS

OSMOTIC pressures have primarily been measured in order to determine molecular weights of proteins and other macromolecules. The history of the method is closely linked with the history of the development of our concepts about solutions of macromolecules. Indeed, in the early stages there was some doubt whether the idea of 'molecular weight' could be used at all, when dealing with so-called 'colloidal solutions'. Such doubts were removed in the late 1920s and 1930s when successively analytical ultra-centrifugation and electrophoresis showed that at least some protein preparations could be regarded as distinct macromolecular entities and a unique 'molecular weight' could validly be assigned. In fact osmotic pressure measurements can claim to have been one of the first methods successfully applied in determining a 'particle' weight.

More recently it has become clear that protein molecules, or 'particles' in solution often consist of parts, the 'sub-units', which internally are linked by covalent bonds but which are joined to each other not by covalent bonds but by a variety of non-covalent links such as hydrogen bonds and 'hydrophobic' and electrostatic interactions. This means that the 'molecular' weight is particularly liable to change as solvent and other conditions are altered. Also, there may be several different molecular species present in a solution of a protein and these may exhibit more or less rapid interconversion. In such a situation the molecular weights of the interacting sub-units, and the average molecular weights of the mixtures are of fundamental importance. Osmotic pressure measurements have been used successfully in investigating this kind of problem, and still continue to be effective with the more complicated systems of current interest, as they were in the earlier determinations of molecular size. The examples described in this chapter are chosen to illustrate the use of osmotic pressure measurements and the various methods of interpreting such observations for a variety of situations.

4.1. Simple protein systems and the precision of osmotic pressure measurements

Haemoglobin

Haemoglobin was the subject of one of the first successful determinations of a protein molecular weight by Adair (1928) and Adair and Robertson

APPLICATIONS OF OSMOTIC PRESSURE MEASUREMENTS

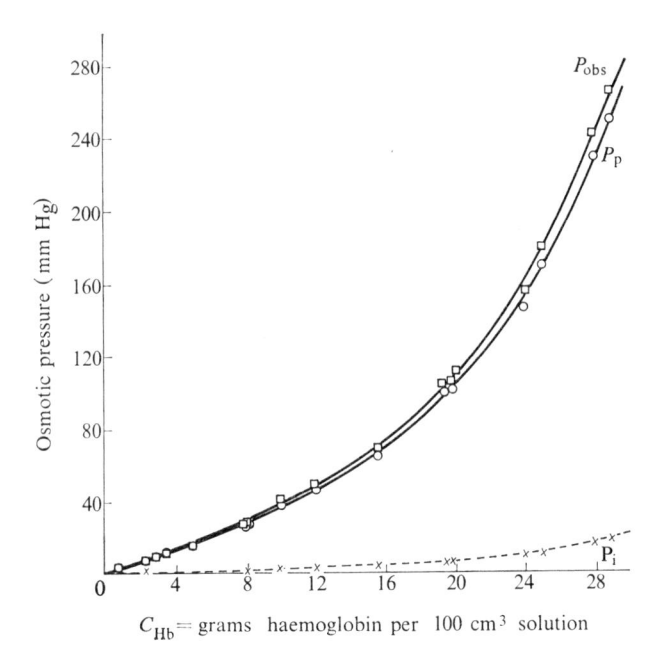

FIG. 4.1. The osmotic pressure of haemoglobin, after Adair (taken from Cohn and Edsall 1943). P_{obs} is the total osmotic pressure in mm (Hg), P_i the calculated Donnan contribution, and P_p that due to the protein. Note the non-linearity of line P_p, showing that non-ideality effects other than the Donnan contribution exist.

(1930). Fig. 4.1 shows a typical set of his very extensive observations in the form of an osmotic pressure–concentration plot. His P_{obs}, which is the total measured osmotic pressure, was corrected for the contribution from diffusible ions, that is, the Donnan effect. This is shown as P_i and gives a good impression of the relative magnitude of this effect, and the way it diminishes to zero at zero protein concentration. The corrected osmotic pressure is denoted as P_P and this plot still shows clear deviations from linearity: hence the solution exhibits non-ideality not due to Donnan effects. It should be noted that the concentration range was extended to the unusually high value of 28 per cent: most workers today would be content to go up to 5 per cent because of the problem of measurement in viscous solutions mentioned in Chapter 3.

It was already known that one gram-atom of iron was contained in about 17 000 g of protein. These observations, which would now probably be presented as a π/c against c plot, showed that the molecular weight of haemoglobin was 67 000 and each molecule contained four atoms of iron.

Haemocyanin and low pressure measurements

Some forty years after the haemoglobin studies just described, Adair & Elliot (1968) determined the molecular weight of the haemocyanin of *Pila leopoldvillensis* with the same simple osmometer that he used for haemoglobin in 1928. The results obtained are shown in Table 4.1. The molecular weight was 8 600 000 in good agreement with the light scattering value, and these results show that, contrary to a common view, molecular weights in excess of 1 000 000 *can* be measured in membrane osmometers. The practical difficulties of simple osmometers have already been discussed at length

TABLE 4.1
Osmotic pressures of haemocyanin of Pila leopoldvillensis in sodium acetate buffers

C g of protein in 100 ml of solution	π osmotic pressure mm toluene at 1°C	Standard error	Maximum error	No. of measure- ments of π
pH 5·20		Ionic strength of buffer 0·15		
0·659	0·207	0·010	0·022	13
1·130	0·416	0·010	0·019	5
1·590	0·590	0·005	0·007	4
1·750	0·749	0·022	0·043	8
2·398	0·948	0·027	0·041	4
pH 5·94		Ionic strength of buffer 0·15		
0·415	0·132	0·009	0·018	8
0·428	0·149	0·007	0·017	13
0·616	0·205	0·009	0·016	5
1·139	0·378	0·011	0·018	5
1·639	0·637	0·010	0·026	59
2·387	1·004	0·007	0·011	4
pH 5·90		Ionic strength of buffer 0·02		
0·509	0·287	0·020	0·047	10
1·019	0·617	0·004	0·014	4
2·008	1·458	0·009	0·015	5

(Section 3.1) and it should be said that sets of measurements were made over a period of four months. Even so, Adair's simple osmometers still have a place when low pressures are expected and it is doubtful if some modern electronic osmometers could have produced such precise results. No doubt the practical skills developed by Adair in forty years' experience of osmometers also contributed.

Heaps & Stainsby (1966) have also used a simple osmometer of a type similar to that used by Adair to examine the low osmotic pressures of solutions of bovine serum albumin (molecular weight $6·9 \times 10^4$), soluble collagen (molecular weight about 3×10^5) and a gelatin (molecular weight 1×10^5) in the concentration range 0·01 to 0·04 per cent protein. As they point out, if very low concentrations are used then the need to set up several osmometers simultaneously to cover the concentration range does not demand unreasonable quantities of protein.

A comparison of two closely-related proteins: ovalbumin and plakalbumin

When ovalbumin is treated with subtilisin a limited proteolysis occurs, and one of the products, plakalbumin, can be crystallized and appears to have

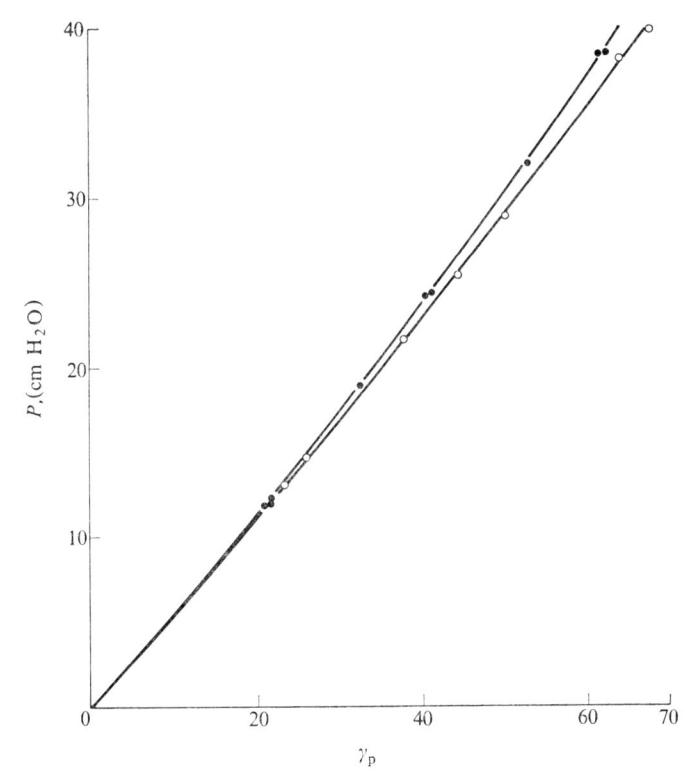

Fig. 4.2. The osmotic pressure of ovalbumin and plakalbumin, taken from Guntelberg and Linderstrom-Lang (1949). P is the total osmotic pressure in cm (H_2O), and γ_p the protein concentration in g l^{-1}. Note the greater curvature of the plakalbumin line (●) as compared with ovalbumin (○).

a high molecular weight. Guntelberg and Linderstrøm-Lang (1949) used osmotic pressure measurements to determine the difference in molecular weight between ovalbumin and plakalbumin and obtained results which are among the most accurate osmotic pressure determination ever made and raise interesting questions on precision. The main observations are shown in Fig. 4.2 and 4.3, and yielded values of 44 900–45 100 for ovalbumin and 44 600–44 800 for plakalbumin. The difference in molecular weight is only about 600: this is actually in good agreement with independent estimates of

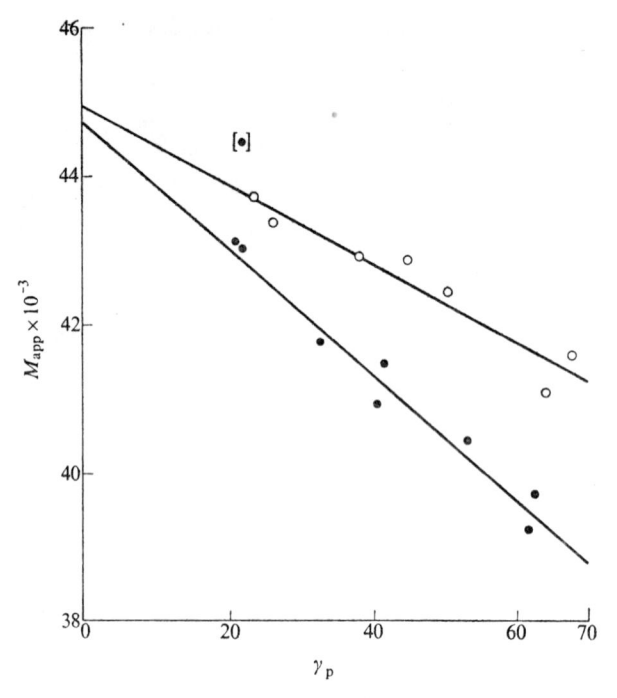

FIG. 4.3. Variation of apparent molecular weight, (calculated from the van't Hoff equation) with protein concentration (g/l) for ovalbumin (○) and plakalbumin (●). The slopes are substantially different. The bracketed point was discarded for the statistical analysis described in Section 4.1.3.

the size of the peptide split off from ovalbumin during the reaction. Plakalbumin always gave lower M_{app} values than ovalbumin, and the slopes of the plots of *M apparent* against concentration (Fig. 4.3) were different.

How significant is this difference in molecular weight? The values are apparently much more precise than would normally be expected. For example the precision of concentration determinations would not normally exceed ±0·5 per cent and no special precautions were taken. In fact Guntelberg employed two methods of analysing the results: in the first he used

$$M_{app} = a + bc_2 \tag{4.1}$$

and by means of the method of least squares he calculated

$$a = \frac{\sum c_2 \sum M_{app} c_2 - \sum c_2^2 \sum M_{app}}{\sum c_2 \sum c_2 - n \sum c_2^2} \tag{4.2}$$

$$b = \frac{\sum c_2 \sum M_{app} - n \sum c_2 M_{app}}{\sum c_2 \sum c_2 - n \sum c_2^2} \tag{4.3}$$

where n is the number of determinations. This is the method employed by most authors (see Section 3.10). It gives equal weight to the various values of M_{app}, but does not allow for the probable greater error at low protein concentrations. The expression

$$c_2 M_{app} = ac_2 + bc_2^2 \tag{4.4}$$

weights the M values according to the values of c_2 and then

$$a = \frac{\sum M_{app} c_2^2 \sum c_2^4 - \sum c_2^3 \sum M_{app} c_2^3}{\sum c_2^4 \sum c_2^2 - n \sum c_2^3 \sum c_2^3} \tag{4.5}$$

$$b = \frac{\sum c_2^2 \sum M_{app} c_2^3 - n \sum M_{app} c_2^2 \sum c_2^3}{\sum c_2^4 \sum c_2^2 - n \sum c_2^3 \sum c_2^3} \tag{4.6}$$

The two methods gave, by the unweighted calculation values for molecular weights as follows: ovalbumin 44 910; plakalbumin 44 760; and, when weighted, ovalbumin 45 070 and plakalbumin 44 630. The difference is not great, which suggests that the precision at low concentration was as good as that at higher protein concentrations. This is theoretically unlikely, but the loss of precision which must occur was presumably so small as to be negligible. Also, since it is the significance of a difference which is at issue, errors in determining protein concentration may have been constant, in the two cases, and are thus eliminated in the comparison. There seems, therefore, no reason to doubt the validity of the difference found, and it appears that osmotic pressure measurements are capable of detecting *differences* in molecular weight of the order of 1·0 per cent.

A quite separate question is the precision of the absolute value obtained for the molecular weight of ovalbumin. X-ray crystallography, or sequence determination techniques give molecular weights much more precisely than osmotic pressures, or any solution method, but are not yet available for ovalbumin so a comparison cannot be made. The most that can be said is that the value of 45 000 agrees well with that obtained by hydrodynamic methods (Fevold 1950).

The question of the relationship between M_{app} and concentration, and the involvement of the virial coefficient was dealt with at length in Chapter 2 because they have such a large influence on the determination of molecular weights. It is obvious from Fig. 4.3 that the difference in slope for ovalbumin and plakalbumin has a determining effect in the final conclusions. What is the significance of this difference? Guntelberg and Linderstrøm-Lang give an interesting and original discussion of the factors influencing the slope. It will be clear from the equation which they used for statistical analysis that only B has been taken into account and that higher coefficients were neglected. If they were significant the plots in Fig. 4.3 would have been curved, and it is possible to apply a statistical test for this (see below). However, Fig. 4.3

shows clearly that within the limits of experimental error, the plot is indeed linear. Hydration, and presumably any difference in hydration, of the two proteins contributes very little to the slope and it was concluded that the slope was principally the result of co-exclusion, and any differences in slope arise from differences in the influence of charge and charge distribution on co-exclusions. Fig. 4.4 shows results collected from several authors; evidently the salt concentration has a large effect on the slope and this cannot all be due to Donnan effects. It is probable that protein–protein associations play a part at high ammonium sulphate levels.

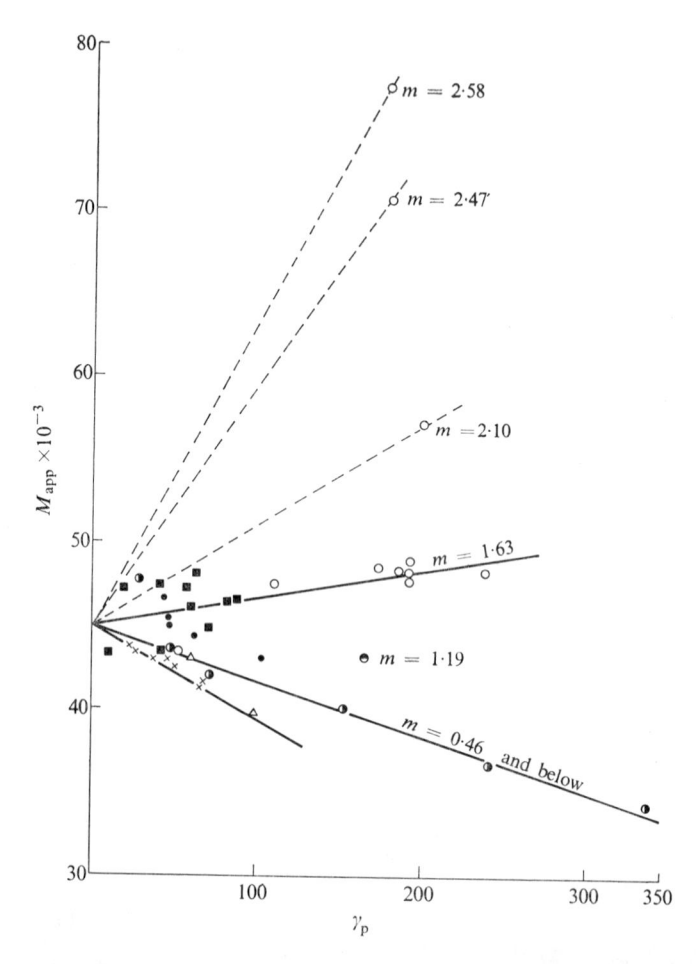

FIG. 4.4. Relationship between apparent molecular weight and ovalbumin concentration (g/l) in different solvents. ○ ◐ ◓ in the indicated molarity of ammonium sulphate. ● in 0·12 M sodium acetate, △ in 1·1 M sodium chloride, ■ and × in sodium phosphate pH 7·0. After Guntelberg and Linderström-Lang (1949).

Thus, even when two closely similar proteins are compared, the virial coefficient can only be accounted for in quite general terms. This is because insufficient is known about the detailed structure of the protein and it would be interesting to examine one of the closely related pairs of haemoglobins or β-lactoglobulin variants for which detailed X-ray crystallographic structures are available (and so precise and unequivocal molecular weights) to see if B can be evaluated in more detail, both absolutely and comparatively.

The precision of an early electronic osmometer

The electronic osmometer devised by Rowe and Abrams (1957) has already been mentioned in Chapter 3. Results obtained with it are included here because these authors applied it to serum albumin and used fairly elaborate statistical tests:

(a) a polynomial regression gave the expression

$$\pi = 0{\cdot}024 + 3{\cdot}665(\pm 0{\cdot}0026)c + 0{\cdot}339(\pm 0{\cdot}001)c^2 - 0{\cdot}0141(\pm 0{\cdot}00037)c^3$$

and molecular weight $= 68\,766 \pm 48$;

(b) a simple linear regression was fitted to π/c against c

$$\pi/c = 3{\cdot}734(\pm 0{\cdot}037) + 0{\cdot}282(\pm 0{\cdot}00049)c$$

and molecular weight $= 67\,500 \pm 670$;

(c) Adair and Robinson (1930) found that c/π against c plots gave good fits, so this was tried and gave

$$10c/\pi = 2{\cdot}652(\pm 0{\cdot}033) - 0{\cdot}159(\pm 0{\cdot}003)c$$

and molecular weight $= 66\,580 \pm 840$.

First, the different methods of treatment produce significantly different results, which must be the result of forcing the observations into three different relationships. In fact, method (a) is clearly the appropriate relationship on theoretical grounds—it takes the third virial coefficient (C) into account, and also yields the lowest standard error.

Because method (a) requires more elaborate calculations, most workers use (b) with some test for linearity. In this instance the linearity in (b) was judged good by inspection, and most workers would regard (b) as an acceptable result. This suggests that for very precise results a polynomial regression should be tried, and that a simple linear regression of π/c vs. c might overlook a small contribution from the third virial coefficient. These authors also make the important point that errors in concentration determination are only partly taken into account, and reasonably claim that no previous work had produced such a small standard error. There would seem to be no theoretical justification for method (c) though this does not detract from its use where it can be demonstrated that it is better than other methods. However, method (a) *should* always give the best fit, for straightforward undissociating systems. It also gives a notably better value for A_1 which would similarly be expected on theoretical grounds.

Heterogeneity in a protein preparation–soluble wool keratin

De Deurwaerder and Harrap (1965) used osmometry and sedimentation equilibrium to investigate a fraction of soluble wool keratin. This is a case where a relatively heterogenous preparation required characterization. The

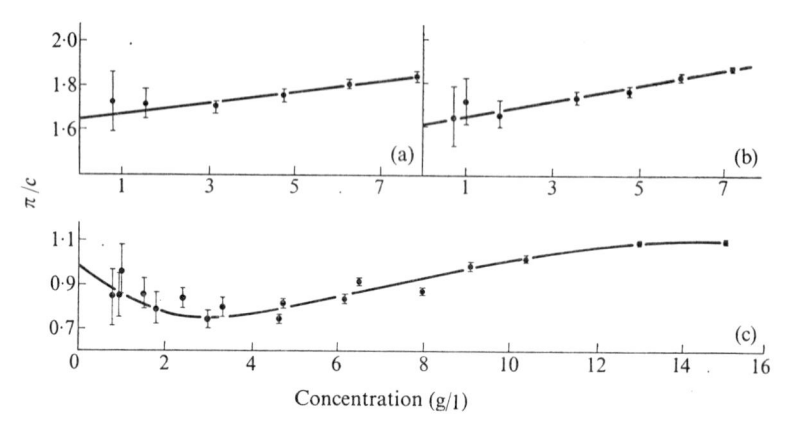

FIG. 4.5. π/c as a function of c for (a) lysozyme; (b) S-carboxymethyl lysozyme; (c) S-carboxymethyl keratin A, a keratin fraction. After De Deurwaerder and Harrap (1965). The main point of interest is the complex shape, and the minimum in the curve in (c).

solvent used was 60 per cent formamide, not an unusual one for this kind of protein.

Some of the results are shown in Fig. 4.5. These authors used lysozyme and S-carboxy methyl lysozyme to check the precision of their osmometer, and obtained a value of 15 100 for the molecular weight of lysozyme, which is known to be 14 307, so the agreement was not good. The lower curve in Fig. 4.5 was obtained with the S-carboxymethylated keratin fraction and plainly shows complicated concentration dependence. Above about 0·4 per cent it could be regarded as normal but below this limit it shows signs of dissociation on dilution. In this kind of situation extrapolation is difficult and the significance of the derived molecular weight is doubtful. Probably the most interesting feature of these results is the existence of a minimum at about 0·3 per cent. It would be interesting to calculate the theoretical relationships for this point where a positive contribution from the virial coefficient is just balanced by a negative contribution from, probably, dissociation but this has not been done. As in the insulin case mentioned below, the solution would behave as if it were ideal. The potential existence of such an inflection can of course be predicted in general terms and this seems to be a good example, as well as illustrating the use of the method on an ill-characterized preparation in a difficult solvent.

4.2. Determination of numbers of sub-units by osmotic pressure measurements

In recent years there has been considerable interest in the quaternary structure of many globular proteins whereby they consist, in their native state, of several peptide chains usually held together by non-covalent forces. The initial questions must be how many such 'sub-units' the protein contains, and how many kinds there are. In order to answer this question, the molecular weight of the protein may be measured when it is intact and when it is dissociated in some solvent such as aqueous guanidinium hydrochloride or urea. A useful survey of such solvents and their relative potency is given by Gordon and Jencke (1963).

In most cases sedimentation techniques have been used and the weight-average molecular weights measured. There are some difficulties with this approach which do not appear in osmotic measurements. First of all, if $(M)_0$ is the molecular weight of the intact protein and M_w the weight-average molecular weight of the sub-units, then n, the number of sub-units, is given by M_0/M_w if, and only if, all the sub-units are identical in molecular weight. Fortunately this usually seems to be the case, but use of M_w requires evidence that this is so. The use of number-average molecular weights for this determination of the number of sub-units requires no such proof, and the necessary relationships have already been given in Section 2.8 (eqn 2.36). These relations have been applied by Kawahara and Tanford (1966) to a protein (rabbit muscle aldolase) for which there was some uncertainty about the identity of all the sub-units and by Tombs and Lowe (1967) to arachin, whose N-terminal amino acids were known to be such that there was more than one kind of sub-unit. Osmometry was deliberately chosen as the most appropriate method in this case. Tanford measured by sedimentation equilibrium runs the M_w (37 000–41 000), M_z (39 000–43 000) and M_n (36 500–40 000) of aldolase in 6 M guanidinium hydrochloride and also the molecular weight of intact aldolase as 158 000 in the absence of guanidinium hydrochloride. Castellino and Barker (1968) measured M_n by osmometry and obtained a value of 42 400 for aldolase in 6 M guanidinium hydrochloride and a value of 156 400 for the intact enzyme.

These results left no doubt that the aldolase molecule contains four similar sub-units, and not three as previously believed by Schachman and Edelstein (1966). The discrepancy almost certainly arose in this earlier work from an erroneous value for $\bar{v}$ in their sedimentation equilibrium calculations. Castellino and Barker made similar measurements on many proteins which are reproduced in Tables 4.2 and 4.3, together with the results of sedimentation and viscosity measurements. This is an important compilation which shows clearly the usefulness of osmotic pressure measurements, especially for difficult solvents, and also shows that results so obtained accord well with other techniques.

The good agreement between the measured values of $[\eta]$ and those calculated from equations for random coil polypeptide chains (Tanford, 1966,

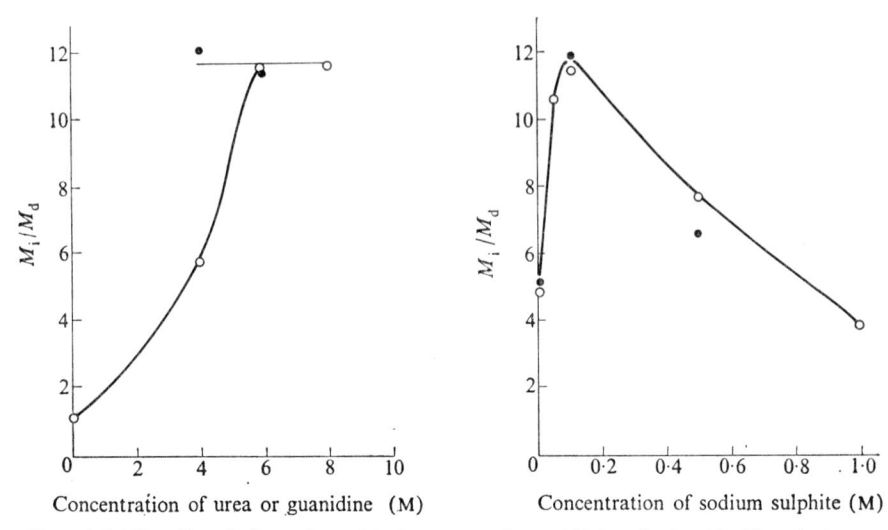

FIG. 4.6. The dissociation of arachin in urea and guanidinium hydrochloride solutions, in the presence of sodium sulphite. M_i is the molecular weight of the native protein, and M_d the determined molecular weight. M_i/M_d is therefore the number of fragments produced on dissociation, and reached a maximum of twelve, indicating the presence of twelve sub-units. From Tombs and Lowe (1967).

1968) suggests that in guanidinium hydrochloride the proteins were approximately in a random coil configuration. This is also reflected in the values for BM_2 which are of the order of 2–5 ml g^{-1} for globular molecules and become greater, rising to values of 50–100 ml g^{-1} for the random coil. The associated decrease in $S_{20,w}$ is due to both unfolding and dissociation to sub-units.

Arachin was believed to contain about 12 sub-units totalling 350 000 g mole^{-1} and it was known that they were not identical. The more numerous the sub-units the more difficult it is to determine their number precisely, and in particular the more important it is that the protein should be fully dissociated. A four sub-unit protein, which is still 10 per cent undissociated in the chosen solvent would still yield to a value of 3·5–4 sub-units from M_n measurements. In a protein of 12 sub-units, the same degree of dissociation would indicate 10 as the apparent number.

TABLE 4.2

The molecular weights of native and dissociated proteins and the number of subunits as determined by osmometry

Protein	Solvent density (g/cm³)	$RT \times 10^{-4}$ (cm l mole⁻¹)	π/c (cm l g⁻¹)	M_n	Subunits ±3%
Serum albumin	1·012	2·4941	0·365±0·003	68 320±600	1·0
Serum albumin+G	1·150	2·0475	0·302±0·005	67 790±1000	
Ovalbumin	1·012	2·4941	0·559±0·003	44 620±300	1·0
Ovalbumin+G	1·150	2·0475	0·440±0·006	46 530±600	
Alcohol dehydrogenase	1·012	2·4941	0·290±0·006	86 000±1750	2·1
Alcohol dehydrogenase+G	1·150	2·0475	0·502±0·003	40 790±300	
Enolase	1·009	2·5015	0·303±0·003	82 550±800	2·3
Enolase+G	1·150	2·0475	0·561±0·003	36 500±200	
Methemoglobin	1·008	2·5040	0·393±0·007	63 720±1100	4·0
Methemoglobin+G	1·150	2·0475	1·293±0·006	15 840±800	
Lactate dehydrogenase	1·012	2·4941	0·183±0·002	136 290±1400	3·8
Lactate dehydrogenase+G	1·150	2·0475	0·566±0·013	36 180±800	
Aldolase	1·008	2·5040	0·160±0·001	156 500±1000	3·7
Aldolase+G	1·150	2·0475	0·483±0·003	42 400±300	

G = proteins in GuHCl–MSH

TABLE 4.3

Values of $s_{20,w}$, $[\eta]$ and B for native and dissociated proteins

Protein	$s_{20;w}$±0·1[a] s⁻¹ (S)	$[\eta]$±1% (cm³ g⁻¹) Observed	$[\eta]$±1% (cm³ g⁻¹) Calculated[b]	$B \times 10^5$±7% (cm³ mole g⁻²)
Serum albumin	4·41	3·6		2·9
Serum albumin+G	1·99	51·3	51·3	92·4
Ovalbumin	3·53	4·4		2·4
Ovalbumin+G	1·39	34·6	35·5	76·0
Alcohol dehydrogenase	5·39	3·6		4·5
Alcohol dehydrogenase+G	1·53	34·1	34·9	77·8
Enolase	5·59	3·7		1·0
Enolase+G	1·53	33·0	33·5	32·4
Methemoglobin	4·40	3·4		1·3
Methemoglobin+G	0·96	19·8	19·7	109·0
Lactate dehydrogenase	7·18	3·8		2·0
Lactate dehydrogenase+G	1·55	32·4	32·1	81·7
Aldolase	7·80	3·4		2·8
Aldolase+G	1·35	35·4	35·6	70·8

[a] Protein concentrations range from 4 to 8 mg per ml. [b] $[\eta] = 0.684 n^{0.67}$, where n is the number of amino acid residues per chain (Tanford *et al.*, 1966).

Fig. 4.6 shows the effect of increasing concentrations of urea or guanidinium hydrochloride on the number-average degree of dissociation of arachin. Such information is always desirable for multi-subunit proteins: in the case of arachin the protein was fully dissociated, so any errors can be arising only in the molecular weight determinations themselves. Values of BM_2 ranged from 27 ml g^{-1} in 8 M urea to 83 ml g^{-1} in 6 M guanidinium hydrochloride, while the intrinsic viscosity in the latter solvent was 35·0 in good agreement with the value expected for a random coil configuration.

A further difficulty with dissociating solvents in sedimentation work is that $\bar{v}$, the partial specific volume of the protein has to be determined.

TABLE 4.4

Comparison of values of $\bar{v}$ for native proteins and values in guanidinium hydrochloride solutions calculated by combining the molecular weight determined by osmometry with sedimentation–equilibrium results. (After Castellino and Barker 1968.)

Protein	Native $\bar{v}$ at 20°	Calculated $\bar{v}$ at 20°
Serum albumin	0·729	0·726±0·002
Ovalbumin	0·744	0·746±0·002
Alcohol dehydrogenase	0·750	0·749±0·001
Enolase	0·747	0·723±0·001
Methemoglobin	0·750	0·746±0·002
Lactate dehydrogenase	0·740	0·732±0·001
Aldolase	0·742	0·743±0·002

There seems to be no reason why it cannot be measured in the usual way in such solvents but small errors can lead to erroneous conclusions. As was shown in 2.7, the value of $\bar{v}$ influences osmotic pressures indirectly, since it appears as a contribution to B, which in its turn affects the extrapolation and hence the molecular weight. It does not need to be measured separately, but is none the less involved. This has been taken to the length of combining M_n from osmotic pressures with sedimentation equilibrium data to obtain apparently accurate values of $\bar{v}$ in 'difficult' solvents (Castellino and Barker 1968, Table 4.4). However, M_n must be equal to M_w for this procedure to be reliable and it hardly seems preferable to direct determinations of $\bar{v}$ from the density of solutions. It could be helpful if there were too severe a shortage of materials for density measurements to be possible.

Lapanje and Tanford (1967) have measured virial coefficients for a series of proteins in 6 M guanidinium hydrochloride–0·1 M mercaptoethanol solutions by osmotic pressures. These measurements along with others showed that, in this solvent, proteins have approximately the shape of random coils.

The idea of the 'random coil' arises from work on less complicated synthetic polymers. It is perhaps surprising that it is only so recently that it was experimentally demonstrated that this concept applies to the conformation of disrupted protein chains. It should be remembered, however, that conformation is dependent on the solvent, and that 'random coil' is a fairly precisely defined concept, which must not be loosely equated with that of 'denatured' protein, without some positive evidence for the identification. Figure 4.7 illustrates for proteins in guanidinium hydrochloride (Lapanje and Tanford

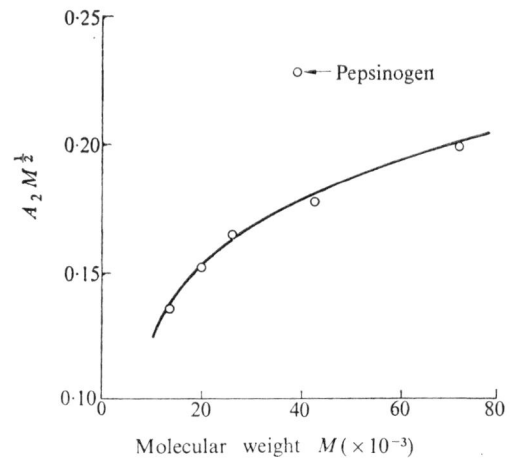

FIG. 4.7. After Lapanje and Tanford (1967). A_2 is a form of the second virial coefficient determined by osmometry for a number of proteins of various molecular weights M. The solvent was 4 M-guanidinium hydrochloride–0·2 M-mercaptoethanol, and the relationship shown suggests that the proteins are in a nearly random coil configuration.

1967) the relationship between the second virial coefficient and the molecular weight and this relationship is, in fact, typical of polymers in random configurations. Proteins do not always give random coils, however, even in this solvent (for example pepsinogen (Fig. 4.7)); the reason for this discrepancy is not known.

4.3. Studies on non-reacting mixtures of like or different proteins

Osmotic pressure measurements on non-reacting mixtures have the advantage that they not only yield a number-average molecular weight, but also a value for the second virial coefficient B as well as interaction coefficients which can give information about the mutual interactions between the component macromolecules. For example, the question whether simple co-exclusions or more complex interactions are occurring may be answered and, by varying the solvent or the protein concentration, some idea of the nature of the interactions may be obtained.

Thus Ogston (1937) employed osmotic pressure measurements to demonstrate the absence of specific interactions in mixtures of serum globulin and albumin. The question had arisen because of the existence of quantitative discrepancies in the apparent proportions of those components which were derived from sedimentation velocity analysis of their mixtures. (This became known as the Johnston–Ogston effect.) Ogston measured the molecular weights, and the dependence of apparent molecular weight on concentration, of albumin and 'serum globulin' both separately and in mixtures of the two. The molecular weights of the mixtures corresponded well with the predicted values, and the dependence of M_{app} on concentration could be explained solely on the basis of co-exclusion effects. Thus specific interactions between albumin and globulin could probably be eliminated as an explanation of the Johnston–Ogston effect and some other explanation then had to be sought, and was successfully found.

Ogston and co-workers (Laurent and Ogston 1963) also used the same approach in the much more difficult case of possible interactions between serum albumin and hyaluronic acid. Their problem was to determine the effect of each solute on the osmotic pressure of the other, or more generally to analyse the concentration dependence of the osmotic pressure initially in terms of a ternary protein–hyaluronic acid–water system. In fact the 'water' component was a salt solution chosen to minimize Donnan equilibria contributions. (This in itself of course means that any interactions discovered are necessarily those in fairly high ionic strengths. Experiments done under such conditions accept this limitation from the start. An example of studies deliberately carried out under conditions of low ionic strength is discussed below.)

The experiments on the serum albumin–hyaluronic acid system were designed to measure the virial coefficients, and to see if these were consistent with the values to be expected from certain postulated types of interaction. It is always possible that interactions could occur which produce quantitatively opposed effects, thereby giving the impression, in the extreme case, that the solution is ideal. This pseudo-ideality occurred with insulin (see Section 4.4.1). Thus, experiments of this kind can only attempt to investigate the internal consistency of results, and make use of the general theoretical background to assess whether or not analysis in terms of a ternary system, instead of more complicated systems is reasonable.

Experience and theory suggested that, in this instance, treating the salt solution as if it were water did not significantly alter the interpretation of the virial coefficients, or their magnitude. This must not be taken to imply that actually varying the ionic strength will have no significant effect, but merely that in these particular conditions and in the light of the results obtained it is not necessary to take salt macromolecule interactions into account to obtain a consistent interpretation of the results. This problem does not arise when molecular weights are the only objective but can hardly be neglected when the virial coefficient is to be subjected to a detailed molecular interpretation.

In this study, the osmotic pressures of separate solutions of albumin and hyaluronic acids were measured and then the osmotic pressures of solutions containing constant amounts of albumin and varying amounts of hyaluronic acid and vice versa. The measured values for the mixture $^{2,3}\pi$ were corrected by the separately measured value for either albumin, $^2\pi$, or hyaluronic acid, $^3\pi$, and extrapolated to either c_2 or $c_3 = 0$ ('2' refers to albumin; '3' refers to hyaluronic acid). The results were

$$\frac{^{2,3}\pi - ^3\pi}{c_2} = 3\cdot 43 \times 10^5 + 8\cdot 55 \times 10^6 c_3 + 3\cdot 17 \times 10^9 c_3^2 \quad (4.7)$$

and

$$\frac{^{2,3}\pi - ^2\pi}{c_3} = 1\cdot 08 \times 10^5 + 8\cdot 4 \times 10^7 c_3. \quad (4.8)$$

The corresponding equations in molal form for a ternary system are (Ogston 1962):

$$\frac{1000}{RT} \frac{^{2,3}\pi - ^3\pi}{m_2} = 1 + a_1 m_3 + a_2 m_3^2 \quad (4.9)$$

$$\frac{1000}{RT} \frac{^{2,3}\pi - ^2\pi}{m_3} = (1 + a_1 m_2) + m_3(\tfrac{1}{2}d + a_2 m_2) \quad (4.10)$$

a_1 and a_2 are forms of virial coefficient, as is d, which is defined by

$$\frac{1000}{RT} \frac{^3\pi}{m_3} = 1 + (\tfrac{1}{2}d)m_3 \quad (4.11)$$

Thus d can be obtained from measurements on solutions of hyaluronic acid alone. Equations 4.9 and 4.10 thus permit two independent estimates of the values of a_1 and a_2. Equation 4.7 gave $a_1 = 1\cdot 01 \times 10^4$ and $a_2 = 1\cdot 57 \times 10^9$, with $M_2 = 7\cdot 1 \times 10^4$ and $M_3 = 4\cdot 13 \times 10^5$. Equation 4.8, together with d gave $a_1 = 0\cdot 62 \times 10^4$ and $a_2 = 1\cdot 1 \times 10^9$. Thus there is reasonably good agreement between the two sets of values, and eqns 4.9 and 4.10 describe the interaction over a wide range of albumin and hyaluronic acid concentrations.

The values of the coefficients a_1 and a_2 may be further interpreted to yield an excluded volume of about 25 ml/g of hyaluronic acid. This is within the range of values obtained from a variety of techniques (Ogston and Preston 1966) and co-exclusion effects are a reasonable basis for the interpretation of these results. Preston, Davis, and Ogston (1965) extended the treatment to include salt, and analysed their results in terms of a salt–water–hyaluronic acid–albumin system. This is probably the most extensive treatment of a mixture yet undertaken, and the equations are daunting. Although they can be written down the point is rapidly reached where experimental precision is insufficient to evaluate the less important dependencies. Disappointingly the much more complex treatment did not yield any fresh insights, and the main conclusion remains unchanged that exclusion effects predominate and that, at least under the conditions of ionic strength used, there are no major specific interactions.

Scatchard and Pigliacampi (1962) deliberately set out to examine charge–charge interactions between serum albumin and carboxy-haemoglobin, and chose accordingly to make measurements in very dilute (0·003 M) sodium chloride solutions.

Some of their results may be summarized as:

m_3	x_2	pH	$M_n \times 10^4$	2β	$2\beta^1$	$2\beta D$
0·003	0·763	5·74	6·83	316	1000	304
0·003	0·502	5·99	6·76	142	3000	82
0·003	0·263	6·61	6·68	386	1000	10
0·01	0·504	5·95	6·76	234	1000	84
0·15	0·524	6·06	6·76	434	0	40

where m_3 is the sodium chloride concentration, x_2 the mole fraction of albumin in the solute mixture of albumin and carboxy-haemoglobin, both of which were isoionic and βD is the contribution to the coefficient from the Donnan effect. βD was calculated from known parameters for the protein. β and β^1 are defined by the equation

$$\pi = \frac{RT}{V^0}\left(1 + \beta M_p - \frac{\beta^1 M_p}{1 + 2\beta^1 M_p}\right) \qquad (4.12)$$

where M_p is the total molal concentration of protein. β^1 is a measure of interaction effects and is very large and dependent on ionic strength. It can thus reasonably be interpreted as a charge–charge interaction effect. Its magnitude calculated from known charge density and molecular sizes was of the correct order of magnitude and general dependence on protein concentration. Theoretical calculations for this kind of interaction are not only very complex but also depend on more detailed knowledge of the distribution and nature of charged groups on the surface of the molecule than are yet available. The effects of electrostatic interactions between macromolecules can be detected as easily by osmometry as by any other method, but in view of the great difficulties of interpretation it is not surprising that few studies of this kind exist. In most cases conditions are chosen to minimize electrostatic effects.

4.4. Associating–dissociating systems

Insulin

One of the earliest pieces of work to take account of the possibility of association of a protein was Gutfreund's (1948) study of insulin. He used the osmometers developed by Adair and the general relationship,

$$\pi - \pi_i = gRTM_2 \tag{4.13}$$

where π_i is the osmotic pressure due to Donnan effects, and g is a form of osmotic coefficient which allows for all the other effects on the second virial coefficient. Many earlier workers preferred to separate out the Donnan effect in this way. Gutfreund points out that g is negligible for protein concentrations below 1 per cent for fairly spherical molecules, and made no further allowance for it. It is interesting that Lauffer (see Section 4.9) has reached a similar conclusion after much more elaborate considerations.

Gutfreund also measured the membrane potential as a check on the magnitude of the Donnan contribution. The molecular weight was 40 300±1400 at 0·5°C but decreased to 32 000±2800 at 20·5°C, both in 0·02 M sodium phosphate pH 7. There was however no variation in determining the molecular weight at concentrations of 0·11 to 0·16 per cent protein, but in the range 0·4 to 0·8 per cent the molecular weight measured had the higher value of 47 700. There was, therefore, some indication of an increase in M_2 with concentration. In a further investigation using π/c_2 against c_2 plots clear evidence of a concentration dependence was obtained. Fig. 4.8 shows some of his observations (Gutfreund 1952) and the upward curvature of the plot is evident in the lower line of Fig. 4.8a. The upward slopes in Fig. 4.8d and 4.8e also suggest clearly that the molecule is dissociating on dilution. Fig. 4.8c is apparently a case of co-exclusion effects balancing out dissociation since the solution behaves as if it were ideal. The plot is flat.

At the time there was some controversy as to the true molecular weight of insulin. It was clearly variable in solution, as judged by a variety of techniques. These results suggested strongly a value of 12 000 at $c_2 = 0$. The molecular weight of the sub-unit is now known to be 6000, by direct determination of the amino acid sequence. It must be presumed that if it had been possible to extend the measurements to lower protein concentrations still, the shape of the curve in Fig. 4.8 would have been different, and a better extrapolation could have been made. It was not possible, and this result cannot be criticized on experimental grounds. The two sub-units of insulin are tightly bound in the solvents chosen and there is no indication from the available results that the molecular weight ever drops below 12 000. It must be emphasized, however, that in such circumstances extrapolation eliminates the various influences on B but does not necessarily yield the molecular weight at zero or very low concentration. B and M_n may vary with concentration to quite different extents in the unknown region. Thus the value of 12 000 is quite correct for the molecular weight of insulin over the concentration range studied, but extrapolation does not automatically ensure that the molecular weight at $c_2 = 0$ is obtained. Later work (Rees and Singer

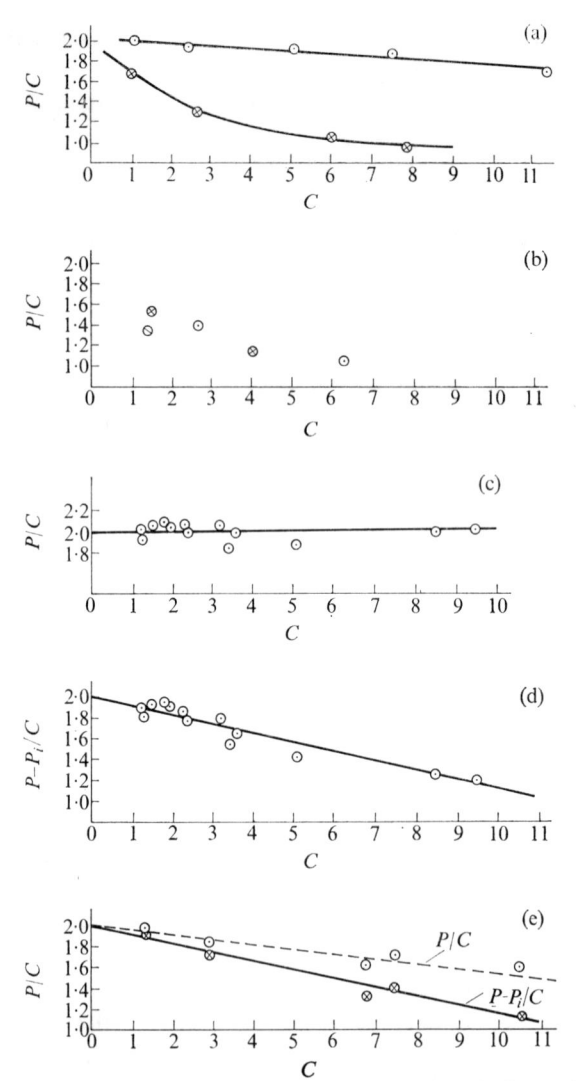

FIG. 4.8. π/c versus c plots for insulin in a variety of phosphate buffers, with and without corrections for Donnan contributions (P_i). Concentrations in g/l and P in cm (H_2O). In every case the slope of the plots indicates dissociation on dilution, even in (c) where the line is apparently flat, and the solution is pseudo-ideal. After Gutfreund (1952).

1955; Kupke and Linderstrøm-Lang 1954) in which guanidinium hydrochloride or dimethylformamide were employed to encourage dissociation has found clear evidence of a species of molecular weight 6000. Complete dissociation of insulin was obtained only in 9 M-guanidinium hydrochloride solution, but the great majority of proteins are completely dissociated at 4–6 M guanidinium hydrochloride, so insulin appears to be an unusual case.

This work has been considered in detail, both because it was the first serious attempt to deal with dissociating protein, and because it illustrates so well the inherent difficulties in extrapolation to eliminate complicated concentration dependence.

Bovine serum albumin

While the main usefulness of the determination of number-average molecular weights is to follow association–dissociation, they also allow estimates of the frequency of interchain links. In a brief study Tombs (1970) has used this to test ideas on aggregation kinetics (Smoluchowski 1918) in the case of heated solutions of bovine serum albumin.

According to Smoluchowski n_0 the initial number of particles is related to n_t, the number at time t by

$$n_t = n_0/1 + 4RD\pi\epsilon n_0 t \qquad (4.14)$$

and since $n_0/n_t = a = M_i/M_n$ where M_i is the initial number average molecular weight and a is the number average degree of aggregation,

$$a = 1 + 4\pi RD\epsilon n_0 t \qquad (4.15)$$

and some estimate of the collision frequency factor ϵ can be made, if R, the radius, and D the diffusion coefficient of the interacting particles can be estimated. Interchain link frequencies can easily be calculated from a, and used to examine the approach to gelation. Little use has been made of molecular weight measurements in work on the aggregation of globular proteins and osmometry might well be fruitfully applied to such systems.

Tobacco mosaic virus protein

Lauffer and co-workers (Banerjee and Lauffer 1966; Paglini and Lauffer 1968; Stauffer, Srinivasan, and Lauffer 1970) have made very extensive studies of the behaviour of tobacco mosaic virus (TMV) protein. Their work includes one of the few examples of osmotic pressure measurements where temperature has been explicitly used as a variable (see also insulin Section 4.7) in order to derive thermodynamic quantities. It also includes the only measurement so far in D_2O.

The most interesting property of TMV protein is that it will assemble, in solution, to structures similar to those in the virus. It is one of the most extreme cases known of the 'self-assembly' properties of proteins. Durham, Finch, and Klug (1971) have summarized the principal intermediates involved. Clearly this is a case where the effects of protein–protein associations are likely to predominate. The theoretical problems involved in dealing with this kind of system were referred to in Chapter 2 and this work is a very interesting example of how the apparently insoluble problems raised there have been resolvable in practice.

In 67 per cent acetic acid TMV protein has a molecular weight of 18 200 and this represents the fundamental unit. In the pH range 6·5–8·0 the molecular weight is about 53 000 at 5°C, but the slope of π/c vs c plots varies with the temperature. Fig. 4.9 illustrates this effect. It is shown as a negative slope, and since other evidence suggests that Donnan effects were small, and in any case produce an opposing slope, a probable explanation for this effect is polymerization of the protein as either the concentration or temperature rises. The polymerizing unit must already be the trimer, and larger polymers still are being formed (perhaps the disc form, with 32 monomer units).

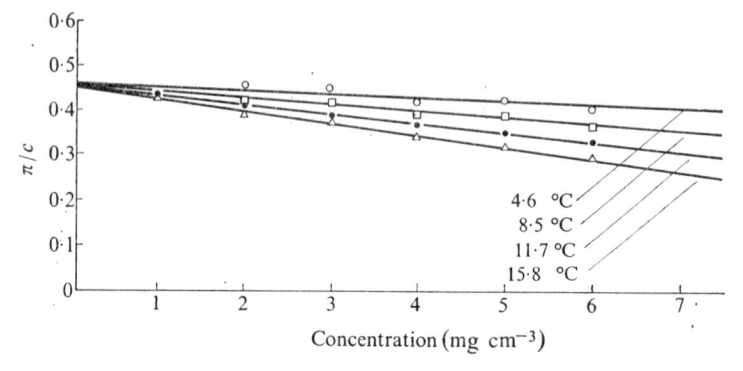

FIG. 4.9. Temperature dependence of the π/c versus c plot for tobacco mosaic virus protein in phosphate buffer $I = 0·067$ pH 7·5. π is in cm (H$_2$O) and c in mg ml^{-1}. Note the change in the slope with temperature, indicating a decrease in dissociation on dilution as the temperature falls. (After Banerjee and Lauffer 1966.)

Fig. 4.10 shows a plot of c against c/π for three different temperatures. This form was chosen because

$$\pi = \frac{(RT)^2}{KM_n}\left(\frac{c}{\pi}\right) - \frac{RT}{K} \qquad (4.16)$$

where K is the equilibrium constant of association at temperature T. Lauffer derived this relationship from condensation polymerization theory and, although it implies ideal behaviour, it can be used for a preliminary analysis. Values for K can be calculated, and from a plot of $\ln K$ against $1/T$, since

$$\ln K = \frac{\Delta S}{R} - \frac{\Delta H}{RT} \qquad (4.17)$$

it was concluded that $\Delta G = 4559$ cal/mole, $\Delta H = 30\,000$ cal/mole and $\Delta S = 124$ eu. Such a large change in the entropy is consistent with a release of about 20 molecules of water for every protein–protein link formed.

Paglini and Lauffer (1968) then attempted the formidable task of direct calculation of both the excluded volume contributions and the Donnan

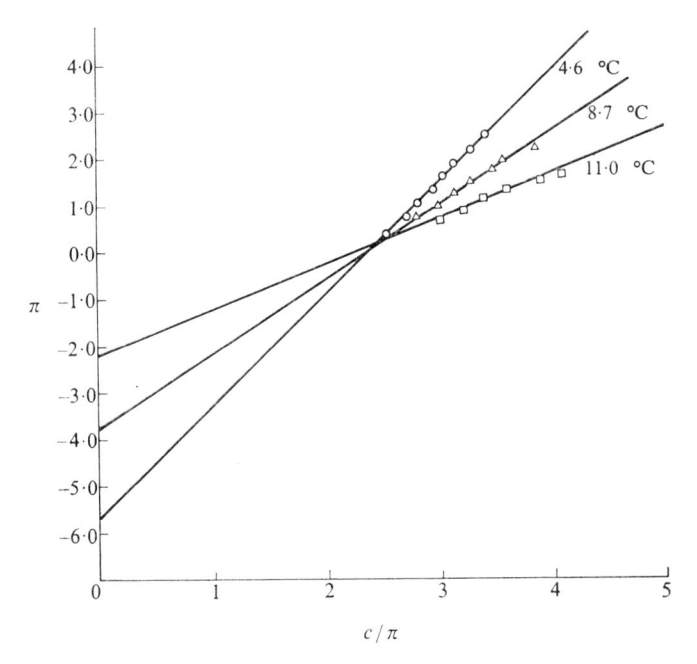

FIG. 4.10. A different way of plotting some of the data in Fig. 4.9. As explained in the text, the slope yields a value for the equilibrium constant of the dissociation, and from its value at different temperatures thermodynamic quantities are available.

effect. They also included a third possible contribution—the effect of polymerization on the extent of hydration of the protein. (This effect, normally very small, will appear in the treatments given above as a complication in calculating the co-exclusion volume and is not usually explicitly taken into account.) Thus, Lauffer defines B (as, in effect does eqn 2.4a) by

$$\pi_0 = \pi_{0i}(1+M_2Bc_2) \tag{4.18}$$

where π_{0i} is the ideal osmotic pressure of the unpolymerized protein of weight M_2, and π_0 is the measured osmotic pressure of unpolymerized protein

$$B = \frac{\xi_0}{M_0^2} + \frac{\beta_{22}^0}{2M_0^2} + \frac{z^2}{4M_0^2 m_3} \tag{4.19}$$

where ξ_0/M_0^2 is a hydration term, which is very small in magnitude, the second term represents co-exclusion effects and the third, the Donnan contribution (see discussion following eqn 2.35 in Chapter 2). In this case all the quantities needed were available in order to calculate that $BM_2 = 0.01$ so that, to a good approximation,

$$\pi_0 = \pi_{0i}(1+0.01c) \tag{4.20}$$

It was known that the Donnan term was not greatly affected by polymerization, and was in any case about twice as large as $\beta_{22}/2M_0^2$.

Exclusion will certainly be affected by polymerization and it is difficult to know how to allow for this. In this instance, it was assumed that exclusion was not dependent on the degree of polymerization significantly. From measured values of π_0 and c, π_{0i} and hence M_n could be calculated, at least partly corrected for non-ideality. The refined value for M_n can then be used to obtain a new value for the equilibrium constant K. This was defined, initially, in concentration terms, but, for finite concentrations, must be corrected for activities. Making assumptions similar to those above,

$$K_{\text{ideal}} = K_{\text{app}}(1+0\cdot0064c). \tag{4.21}$$

Thus, measured values of osmotic pressure are corrected to π_{0i}, which yields M_n. This leads to K_{app}, which is further corrected to K_{ideal}. K_{ideal} was shown to be reasonably constant over a range of concentrations, which is a fairly good test of the model. The corrections actually produced little change in the thermodynamic parameters, and in this case at least were quantitatively small.

This appears to be the only attempt so far to deal fully with polymerizing systems, and was clearly a favourable case since so much information was already available, and a relatively uncomplicated condensation polymerization was involved. More complex situations, with several interacting species remain intractable. This work does suggest, however, that problems which in a theoretical analysis loom large may become less significant when quantities can be assigned to them. We are left with the platitude that ancillary data, and other experimental approaches are always desirable and too much reliance cannot be placed on one technique.

The results in D_2O were different from those in H_2O, which perhaps might be expected since water is intimately involved with protein–protein interactions in general and TMV protein in particular. Both the detailed effects and apparently the type of polymerization were different; the extent of association was generally less in D_2O. This is the opposite of the usual effect of D_2O, which enhances association in glutamate and lactate dehydrogenase (Lee and Berns 1968) and in arachin (Tombs unpublished).

Soyabean proteinase inhibitor

Harry and Steiner (1969) have studied the self-association of soyabean proteinase inhibitor by using osmometry. They assumed that the solutions were ideal, and calculated M_n from the Van't Hoff equation.

$$M_n = \frac{RT}{\pi}c_2 \tag{4.22}$$

The procedures of Section 2.8 (eqns 2.42 to 2.46) were then applied with the results shown in Table 4.5.

TABLE 4.5
Dimerization of soyabean proteinase inhibitor

pH	Temperature	Ionic strength	$K_2 \times 10^{-3}$	ΔG (kcal)
7	25°C	0·1	3·65	−4·85
7	25	1·0	7·97	−5·32
6	25	0·1	10·6	−5·48
5	25	0·1	17·3	−5·77
5	5	0·1	18·6	−5·41
3	25	0·1	5·14	−5·06

K_2 = dimerization constant

The ionic strength was always fairly high so that major non-ideality effects due to Donnan equilibria were avoided. The protein concentration was low (less than 3 mg/ml), so the assumption of ideality probably did not introduce serious errors. Within the limit of experimental precision, the assumption of a monomer-dimer process appeared to be justified. The association was enhanced at high ionic strength, and this is evidence that charge–charge repulsions play a part in preventing association. The effect of pH was similarly related to charge since a shift of pH away from the isoelectric point (4.2) in either direction decreased association.

From the determined charge on the protein and its approximate radius (derived from the molecular weight), Harry and Steiner calculated the difference in electrostatic free energy which results from association. This agreed surprisingly well with the change in ΔG caused by altering the various conditions, with the exception of a shift from pH 6 to 7. There can be little doubt that electrostatic interactions are the main effects controlling the association. Previously, when discussing mixtures it was pointed out that Scatchard *et al.* deliberately sought out conditions where electrostatic effects predominated, and that these were at very low ionic strengths. Some proteins can evidently show large charge–charge interactions at ionic strengths which would usually be regarded as sufficient to suppress them. In fact even such an elementary property as solubility shows that the effect of ionic strength variation on proteins in solution is extremely complex and there are no simple rules.

Phycocyanin

Berns (1970) made some measurements on phycocyanin which is a protein known to show association–dissociation behaviour. In sodium phosphate buffers ionic strength of 0·10 he found the results given in Table 4.6 and after correlation with concurrent ultracentrifuge experiments concluded that these indicate a 'trimer' of weight 90 000 and 'hexamer' of 180 000. The most remarkable result was that π/c against c plots not only showed no evidence at all of dissociation on dilution but also virtually ideal behaviour i.e. $B = 0$.

TABLE 4.6
Molecular weight of phycocyanin

pH	Temperature °C	M_n
6·0	13·7	164 000
6·0	29·4	198 000
7·0	14·3	142 000
7·0	5·3	89 400

There is ancillary evidence from gel chromatography that phycocyanin only shows concentration dependent association at very low concentrations, well below those used in osmometry, though evidently it is strongly temperature dependent. This is still not sufficient to account for why B was zero, and it is possible that a small degree of dissociation just balanced the normal positive contribution of exclusion to B (*cf.* insulin).

Hybridization of haemoglobin

Much more recently osmotic pressure measurements have again been applied fruitfully to an examination of the dissociation of haemoglobins (Guidotti 1967). Haemoglobin dissociates at high dilutions to give dimers according to:

$$\alpha_2\beta_2 \rightleftharpoons 2\alpha\beta, \tag{4.23}$$

When a mixture of haemoglobins is allowed to dissociate

$$(\alpha_2\beta_2)^J + (\alpha_2\beta_2)^K \rightleftharpoons 2(\alpha\beta)^J + 2(\alpha\beta)^K, \tag{4.24}$$

and it is possible that the species $(\alpha\beta)^J(\alpha\beta)^K$ could also form. This question is of interest because the mechanism of such hybridizations could throw light on the nature of the interactions between the sub-units of haemoglobin. It has not been easy to demonstrate the existence of hybrids. Attempts to isolate the hybrid or to show the presence of three four-chain forms had failed. This was probably because all the methods, such as electrophoresis, had to be used under conditions in which the protein was predominantly in the four-chain form, and the pure species, which are the only ones detected, are much more stable than the hybrid. There are further complications because most analytical methods involve mass transport, with continuous re-equilibration, and one could fail to detect hybrids if they converted back to the pure species faster than they were separated during fractionation. What was required, therefore, was some method of directly examining solutions without fractionating in any way.

Suppose that the pure species J at some concentration c_J has a particular number-average molecular weight Mn_J, and that we further determine the variation of Mn_J with concentration c_J. The same can be done for species K. From the total concentration $c_J + c_K$ the number-average molecular weight of mixtures can then be predicted, providing that there is no interaction. If hybrids do form, the molecular weight will always be higher than it would otherwise be, and the difference between the measured and predicted molecular weights can be used to estimate the extent of hybridization. Osmotic measurements are particularly well suited to analysing this situation.

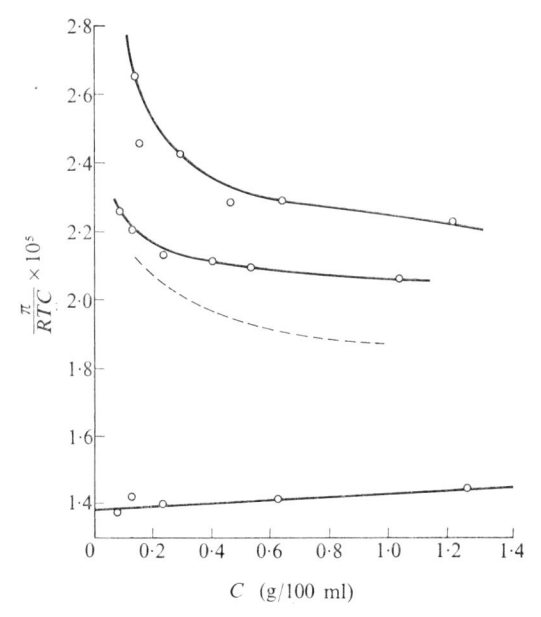

FIG. 4.11. Osmotic pressure results from two proteins and a mixture. The uppermost curve is for oxyhaemoglobin, the bottom one for bovine serum albumin and the central one for an equimolar mixture. The dashed line is a calculated theoretical one; note that it lies below the measured line. The shape of the oxyhaemoglobin curve suggests extensive dissociation on dilution, which still occurs in the mixture (After Guidotti 1967).

The osmotic pressure of a non-hybridizing mixture is approximately given by (cf. eqn 2.4a):

$$\frac{\pi}{RTc} = \frac{1}{M_T}(1+y) + Bc \qquad (4.25)$$

where y is the fraction of the total present in the $\alpha\beta$ form, and c is the total concentration of all species. M_T is the molecular weight of the $\alpha_2\beta_2$ form.

This equation is approximate since a single virial coefficient is not adequate to describe a mixture of species. At least one cross coefficient is needed: however these are positive so that the predicted line in a π/c vs. c plot calculated from this equation must always lie below the measured correct line. This is illustrated in Fig. 4.11, where the discrepancy for an albumin–haemoglobin mixture must be mainly due to failure to use the correct set of B values. The measured line lies above that predicted on the basis of eqn 4.25.

130 APPLICATIONS OF OSMOTIC PRESSURE MEASUREMENTS

Guidotti shows several sets of observations, of which Fig. 4.12 includes representative ones, showing that for mixtures of haemoglobins the measured line always lies below that predicted. This discrepancy cannot be explained by mutual exclusion effects, between the different macromolecular species and by far the most likely explanation is that it results from the formation of hybrids. This approach is quite general and could always be used to demonstrate specific interactions in mixtures.

By fairly obvious extensions, for which Guidotti derives the appropriate equations, it is possible to estimate the values of the equilibrium constants

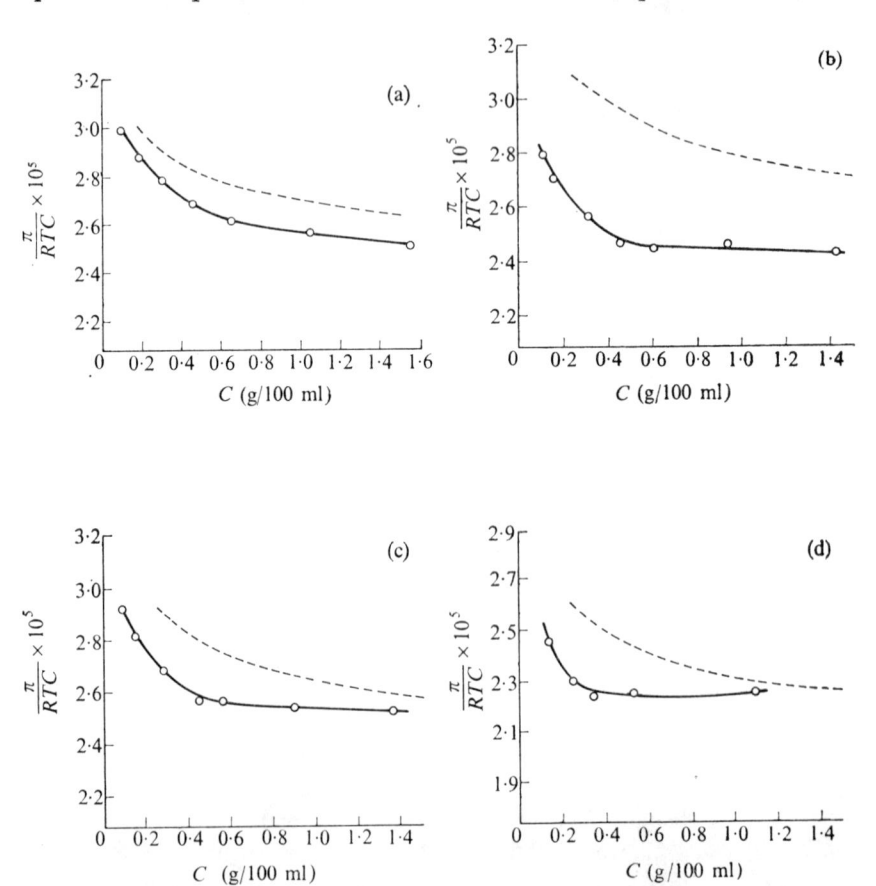

FIG. 4.12. Collected results for mixtures of various haemoglobins. Note now, that in contrast to Fig. 4.11, the measured curve always lies below the calculated curve (dashed line), though dissociation on dilution is still occurring in every case. This is evidence for hybrid formation between the components of the mixture. (a) Carboxyhaemoglobin and iodoacetamide-treated cyan-methaemoglobin in 0·4 M $MgCl_2$ pH 7. (b) As in (a) with, in addition, N-ethyl-maleimide-treated cyanmethaemoglobin and carboxyhaemoglobin. (c) carboxyhaemoglobin, iodoacetamide–carboxyhaemoglobin. (d) carboxyhaemoglobin, iodoacetamide–carboxyhaemoglobin and N-ethyl-maleimide–carboxyhaemoglobin in 2 M NaCl pH 7.

of the dissociations. With haemoglobins, which are mostly of similar size, simplifying assumptions about the similarity of B for different species are justified and make such estimates reasonably precise, but this could not be assumed for the general case. Johnson and Perrella (1971) have further vindicated the osmotic method by using an Adair-type osmometer to determine the dissociation constant of tetramers of sheep haemoglobin into dimers at neutral pH and 0·1 ionic strength and also the corresponding enthalpy change. Gilbert (1966) has shown the presence of hybrids by an ingenious gel chromatography method and this may be a preferable approach.

Histones

Histones are small basic proteins (of molecular weight of the order of 10^4) which contain relatively high proportions of the basic amino acids arginine and lysine and which constitute the principal proteins of nuclear chromatin, where they are complexed with DNA. They can be isolated as chemically identifiable fractions which, surprisingly, vary little from one organism to another. Although they have a high positive charge at neutral pH, some histones have a marked tendency to aggregate, even in solutions containing reasonable amounts of neutral salt, and this has, apart from the case of histone† F1 (Teller, Kinkade, and Cole 1965; Haydon and Peacocke 1968), confused the determination of their molecular weights (*cf.* Edwards and Shooter 1969). This confusion has been compounded (even in the case of F1) by their possession, in spite of their small size, of a very high virial coefficient on account of their large charge (Haydon and Peacocke 1968a) which renders difficult any simple analysis of the variation with concentration of their apparent M_{wc} and M_{nc} values (see Section 2.8). However, this association has acquired a new significance as the result of suggestions (e.g. Crick 1971) that interactions between the histone moieties of DNA-histone complexes may be important in determining the structure of chromosomes and so of their functioning; and because of evidence from nuclear magnetic resonance studies that such interactions are structurally significant in solution (Boublik, Bradbury, and Crane-Robinson 1970; Boublik, Bradbury, Crane-Robinson, and Johns 1971).

Osmotic pressure measurements should be particularly suitable for the examination of this kind of system (q.v. Section 2.8). In an earlier study (Luck *et al.* 1956), osmotic pressure measurements were made to determine molecular weights in guanidinium hydrochloride solution but the histone fractions were not as pure as those more recently available. A preliminary osmotic pressure study (Diggle and Peacocke 1968) on acid-extracted 'histone IIb' (which subsequently turned out to be a mixture of histones† F2a2 and F2c (Diggle and Peacocke 1971)) yielded plots of π/c vs. c some of which passed

† Using the nomenclature of Butler, Johns, and Phillips, 1968.

132 APPLICATIONS OF OSMOTIC PRESSURE MEASUREMENTS

through minima and some through maxima, and so exemplified a situation in which association–dissociation equilibria and a large second virial coefficient were simultaneously operative. A fuller investigation was therefore warranted on purified preparations of chicken erythrocyte histones and some of the results of the osmotic pressure studies on histone F2b are presented (as π/c vs. c) in Fig. 4.13: similar results, with somewhat flatter minima, were obtained at an ionic strength of 0·1 with histones F2a2 (at pH 2 to 7·5) and F3 (at pH 2 and 3). (The osmometer used was that manufactured by Mechrolab, Section 3.5.) The molecular weights of the smallest associating units were determined by measuring osmotic pressures in 6 M guanidinium hydrochloride, when the plots of π/c vs. c were linear and of small (or zero) slope (*cf.* Diggle and Peacocke 1968, Fig. 2). The results are presented in Table 4.7

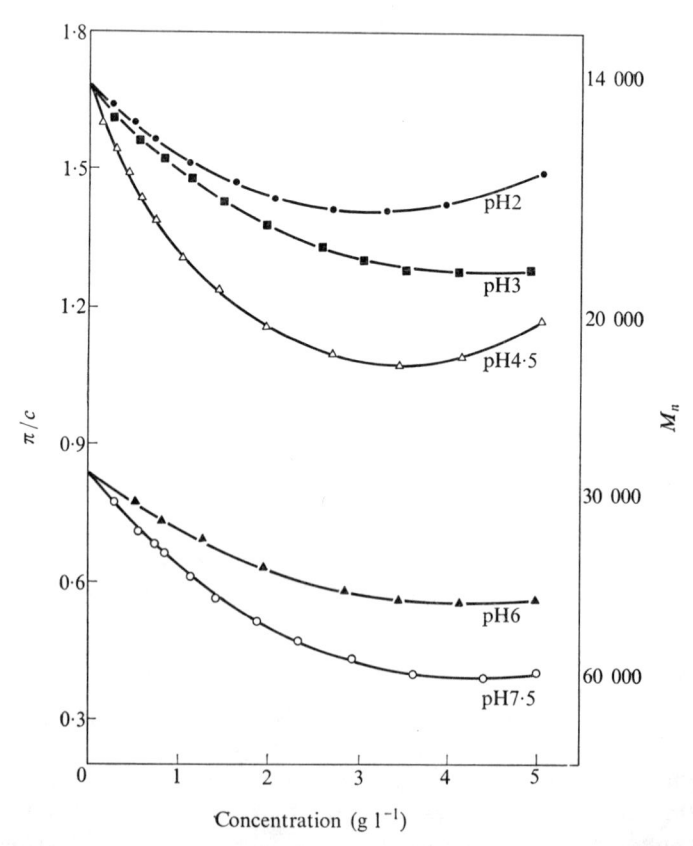

FIG. 4.13. π/c versus c relationship for histone F2b, $I = 0·1$ at the pH's indicated. To the right is shown an M_n apparent scale. The curves are clearly complex, suggesting nonideality and possible dissociation but also show minima. This figure is further discussed in the text (Section 4.4).

APPLICATIONS OF OSMOTIC PRESSURE MEASUREMENTS

TABLE 4.7
Molecular weights of histones in 6 M guanidinium hydrochloride, 25°C

Source: Fraction	Chicken erythrocytes M_n	M_w	Calf thymus M_n	M_w
F2c	20 800±500	20 900±1000	—	—
F2a1	12 300±400	13 000±800	{12 400±500 {†12 500±500	{12 900±1000 {†12 900±1000
F2a2	16 800±500	16 400±1000	{16 700±500 {‡15 700±500	{16 700±1000 {‡16 400±1000
F2b	14 300±400	14 600±1000	14 800±500	14 200±1000
F3	18 600±500	19 600±1000	18 500±500	19 200±1000

Results of Diggle and Peacocke (1971) on histones prepared by the methods of Butler, Johns, and Phillips (1968) and of Johns and Diggle (1969). Fraction (†)GAR = F2a1 and (‡)fraction AL = F2a2 were both provided (in 1967) by Prof. W. C. Starbuck of the Texas Medical Centre, Houston, U.S.A. M_n = molecular weight by osmotic pressure. M_w = molecular weight by sedimentation-equilibrium.

where the values are compared with M_w of the same fractions in 6 M guanidinium hydrochloride and also with M_n in the same solvent of the corresponding fraction isolated from calf thymus. Histones F2a1 and F3 exhibited a very great tendency indeed to aggregate (to species of molecular weight ~10^6 or more) and the measurements of minimum 'monomer size' in guanidinium hydrochloride were especially valuable in these instances. The values of π/c at various concentrations for histones F2b, F2a2 and F3 yielded directly values of M_{nc}^{app} as a function of c and so of M_1/M_{nc}^{app}, since M_1 could be derived either from the measurements of the histones in 6 M guanidinium hydrochloride or, in a few cases, by extrapolation to zero concentration of the M_{nc}^{app} values in a particular solvent. A polynomial expression of the form

$$\frac{M_1}{M_{nc}^{app}} = 1 + Ac + 3c^2 + Cc^3 + Dc^4 + \ldots$$

was derived by computer fitting from these values for each histone in each solvent and thence the corresponding values of M_{wc} could be computed from the slopes of M_1/M_{nc}^{app} vs. ln c, by application of eqn (2.50b). Thereafter the procedures of Adams were followed as outlined in Section 2.8, eqns (2.52). By applying his criteria, it could be shown that each set of observations of π at various c in a particular solvent could be adequately and consistently accounted for if the systems were of the monomer-dimer type with no further aggregation, in solutions of histones F2b and F2a2 at pH 2 to 7·5 and F3 at pH 2 and 3 (all at ionic strength 0·1). However the entity dimerizing at higher pH, was often already a dimer of the smallest 'monomer' or a dimer whose formation could be observed with increasing c at lower pHs. This is indicated in the table of results (4.8) by denoting as H the smallest monomer unit of the particular histone. The association of these histones, under the above

TABLE 4.8

The association of histones at ionic strength 0·1 25°C

	pH	Association system†	Association constant K(dl/g)	$\Delta G°$ of association (kcals/mole)	BM_1 (dl/g)	Charge per molecule of H
Histone F2b	2·0	$2H \rightleftharpoons H_2$	1·8±0·4	−13·6	0·36±0·06	+10
	3·0	$2H \rightleftharpoons H_2$	4·2±0·8	−14·4	0·66±0·09	+14
	4·5	$2H \rightleftharpoons H_2$	28·1±2	−16·3	1·1 ±0·1	+18
	6·0	$2H_2 \rightleftharpoons H_4$	6·4±1	−15·0	1·2 ±0·15	+26
	7·5	$2H_2 \rightleftharpoons H_4$	29·2±4	−17·0	1·7 ±0·2	+31
Histone F2a2	2·0	$2H_2 \rightleftharpoons H_4$	3·8±0·7	−15·1	0·52±0·1	+13
	3·0	$2H_2 \rightleftharpoons H_4$	11·0±2	−16·2	1·1 ±0·1	+19
	4·5	$2H_2 \rightleftharpoons H_4$	25·0±7	−17·0	1·4 ±0·15	+21
	6·0	$2H_4 \rightleftharpoons H_8$	15·0±4	−17·2	1·8 ±0·15	+25
	7·5	$2H_4 \rightleftharpoons H_8$	25·9±6	−17·8	1·8 ±0·15	+25
Histone F3	2·0	$2H_2 \rightleftharpoons H_4$	2·5±0·7	−14·9	0·9 ±0·1	+18
	3·0	$2H_2 \rightleftharpoons H_4$	24·3±4	−17·1	0·8 ±0·1	+18

† The monomer units, H, are 14 400, 16 800, and 19 100 for histones F2b, F2a2, and F3, respectively. H_n is an n-mer of H. The K and $\Delta G°$ refer to the association shown.

conditions, was not of the 'infinite' type since this would have predicted, contrary to observation, a maximum in the M_1/M_{wc}^{app} vs. c plot near to $c = 0$, when BM_1 is positive (as it is for histones).

The procedure of Adams also yielded, as it must if it is to be successfully applied, values of the virial coefficient, from which an estimate of the effective (positive) charge on the histone molecules could be made (Haydon and Peacocke 1968b; Casassa and Eisenberg 1964): these are given in the last column of Table 4.8. The values of BM_1 and of the charge are consistent both with the known hydrogen ion dissociation behaviour of these proteins and with values similarly calculated from parallel sedimentation equilibrium studies (Diggle and Peacocke 1971), which also yielded the same diagnosis of dimerization and values of the dimerization constants very close to those obtained from the osmotic pressure studies (Table 4.8).

4.5. Conclusion

Osmotic pressure measurements have continued to make a major contribution as the interests of physical biochemists have progressed from a simple determination of molecular weight to a more detailed interest in interactions and the effects of solvents on the conformation and integrity of biological macromolecules. The constant underlying thread running through all the examples is the analysis of non-ideality effects and attempts to interpret these in molecular terms. That the same is true of the general study of macromolecules in solution merely emphasizes the close connection between osmotic pressure techniques and other solution methods, and their joint dependence on a fully developed theory of solutions.

Many difficulties remain to be resolved but a great deal of useful information has been obtained from osmotic pressure measurements and there is no reason to suppose that the technique cannot be developed further. Osmotic pressure measurements can readily be made, though considerable care is still required to obtain reliable values, and the results can be interpreted with confidence, in relation to many problems of current interest. The method remains as one of the major techniques for analysing solutions of macromolecules.

REFERENCES

Numbers in square brackets denote the pages on which the reference will be found.

ADAIR, G. S. (1925) *Proc. R. Soc. A.* **108**, 627. [2, 43, 69]
—— (1928) *Proc. R. Soc. A.* **120**, 595, 573. [2, 43, 104]
—— (1961) in *Analytical methods of protein chemistry*, ed. P. Alexander and R. J. Block, Pergamon Press, Oxford. [2, 43, 67]
—— and ELLIOTT, F. G. (1968) *Nature* **219**, 81. [105]
—— and ROBINSON, M. E. (1930) *Biochem. J.* **24**, 1864. [104]
ADAMS, E. (1965a) *Biochemistry* **4**, 1646. [60]
—— (1965b) *Biochemistry* **4**, 1655. [60, 62]
—— (1967) Fractions No. 3 p. 1. Beckman Instruments, Palo Alto, California. [61]
—— and FUJITA, H. (1963) *Ultracentrifugal analysis* (ed. J. W. Williams) Academic Press, New York and London. [60]
—— and LEWIS, M. S. (1968) *Biochemistry* **7**, 1044. [61]
BANERJEE, K., and LAUFFER, M. A. (1966). *Biochemistry*, **5**, 1957. [123]
BERNS, D. J. (1970) *Biochem. Biophys. Res. Comms.*, **38**, 65. [127]
BONNER, R. V., DIMBAT, M., and STROSS, G. H. (1958). *Number average molecular weights*. Interscience, New York and London. [2]
BOUBLIK, M., BRADBURY, E. M., and CRANE-ROBINSON, C. (1970). *Europ. J. Biochem.* **84**, 486. [131]
—— and JOHNS (1970). *ibid.* **17**, 151. [131]
BUTLER, J. A. V., JOHNS, E. W., and PHILLIPS, D. M. (1968). *Prog. Biophys. molec. Biol.* **18**, 211. [131]
CALDIN, E. F. (1958). In *Chemical thermodynamics*, p. 369ff. Clarendon Press, Oxford. [10]
CASSASSA, E. F. and EISENBERG, H. (1960). *J. phys. Chem.* **64**, 753. [3, 19, 20]
—— —— (1964). *Adv. Prot. Chem.* **19**, 287. [3, 19, 20, 38, 134]
CASTELLINO, F. J. and BARKER, R. (1968). *Biochemistry* **7**, 2207. [55, 113, 116]
CHARLESBY, A. (1954). *Proc. R. Soc. A.* **224**, 120. [56]
CLAESSON, S. and JACOBSSON, H. (1954). *Acta chem. scand.* **8**, 1843. [86, 87]
COHN, E. J. and EDSALL, J. T. (1943). *Proteins, amino acids and peptides*. Reinhold New York. [2]
CRAIG, L. C., KONIGSBERG, W., STRATHER, A., and KING, T. P. (1957). *Symposium on protein structure* (ed. A. Neuberger). Methuen, London. [90]
CRICK, F. H. C. (1971). *Nature (Lond.)* **234**, 174. [131]
DAVIES, M. (1966). *Makro. Chem.* **90**, 91, 108. [75, 79]
DERECHIN, M. (1968). *Biochemistry* **7**, 3253. [61]
—— (1969a) *Biochemistry* **8**, 921. [61]
—— (1969b) *Biochemistry* **8**, 927. [61]
DE DEURWAERDER, R. and HARRAP, B. S. (1965). *Makro. Chem.* **86**, 98. [112]

REFERENCES

DIGGLE, J. H. and PEACOCKE, A. R. (1968) *F.E.B.S. Letts*, **1**, 329 (1971). (a) F.E.B.S. Letters, **18**, 138; (b) Unpublished observations (see *also* J. Diggle, D.Phil. thesis, Oxford University, 1971). [62, 131]

DONNAN, F. G. (1911). *Z. Elektrochem*. **17**, 572. [2, 17]

—— (1935). *Trans. Farad. Soc*. **31**, 80. [17, 22]

—— and BARKER, A. (1911). *Proc. R. Soc. A*. **85**, 557. [17]

DURHAM, A. C. H., FINCH, J. T., and KLUG, A. (1971). *Nature* **229**, 37. [123]

EDMOND, E., FARQUHAR, S., DUNSTONE, J. R., and OGSTON, A. G. (1968). *Biochem J*. **108**, 775. [88]

—— and OGSTON, A. G. (1968). *Biochem. J*. **109**, 569. [52]

EDSALL, J. T., EDELHOCK, H., LONTIE, R., and MORRISON, P. E., (1950). *J. amer. Chem. Soc*. **72**, 4641. [18, 48]

—— and COHN, E. J. (1943). *Proteins, amino acids and peptides*, p 382. Reinhold, New York. [55]

EDWARDS, P. A. and SHOOTER, K. V. (1969). *Biochem. J*. **114**, 227. [131]

EINSTEIN, A. (1905). *Ann. Physik*. **17**, 549. [28]

ENOKSSON, B. (1948). *J. Polymer. Sci*. **3**, 314. [85]

FEVOLD, H. L. (1950) *Adv. Prot. Chem*. **6**, 193. [109]

FLORY, P. J. (1953). *Principles of polymer chemistry*. Cornell University Press. Ithaca. New York. [2, 52]

FRAZER, J. C. M. (1931). *A treatise on physical chemistry*. ed. H. S. Taylor, 2nd edition. p 353. Van Nostrand, New York. [2]

FUOSS, R. M. and MEAD, D. J. (1943). *J. Phys. Chem*. **47**, 59. [71, 73]

GIBBS, J. W. (1876). *Collected Works*. vol. I. Longmans, Green & Co. (1928), London and New York. [2]

GILBERT, G. A. (1966). *Nature* **212**, 296. [131]

GLASSTONE, S. (1948). *Textbook of physical chemistry*. McGraw-Hill, New York. [2, 36]

GORDON, J. A. and JENCKE, W. P. (1963). *Biochemistry* **2**, 47. [113]

GROSS, L. M. and STRAUSS, U. P. (1966). In *Chemical physics in ionic solutions* (ed. B. F. Conway and R. G. Barradas), p. 361. Wiley, New York. [54]

GUGGENHEIM, E. A. (1967). *Thermodynamics* (5th edit.) North-Holland Publishing Co., Amsterdam. [57]

GUIDOTTI, G., (1967). *J. Biol. Chem*. **242**, 3694. [128]

GUNTELBERG, A. V. and LINDERSTRØM-LANG, K. (1949). *C.r. trav. Lab. Carlsberg. Ser. Chem*. **27**, 1. [74, 107]

GUTFREUND, H. (1948). *Biochem. J*. **42**, 156, 544. [54, 120]

—— (1950) *Progr. Biophysics*. **1**, 1. [2]

—— (1952) *Biochem. J*. **50**, 566. [54, 121]

HANSON, A. T. (1961). *Acta physiol. Scand*. **53**, 197. [82]

HARRY, J. B. and STEINER, R. F. (1970). *Biochemistry* **8**, 5060. [126]

HAYDON, A. J. and PEACOCKE, A. R. (1968a) *Chemical Society Special Publication No.* 23, 315. [62, 131]

—— —— (1968b). *Biochem. J*. **110**, 243. [62, 131, 134]

HEAPS, P. W. and STAINSBY, G. (1966). *Unpublished observations* (also Stainsby, G., Ph.D. thesis, University of Leeds, 1966). [106]

HILL, T. L. (1956). *J. amer. Chem. Soc*. **78**, 4281. [11]

HILL, T. L. (1958). *J. am. Chem. Soc.* **80**, 2923. [11]
—— (1960). *An introduction to statistical thermodynamics.* Addison Wesley, Reading, U.S.A., and London, England. [11]
JACOBSSON, G. (1954). *Acta Chem. Scand.* **8**, 1843. [86]
JOHNS, E. W. and DIGGLE, J. H. (1969). *Eur. J. Biochem.* **11**, 495. [133]
JOHNSON, J. A. and WILSON, T. A. (1967). *J. Theoret. Biol.* **17**, 304. [63]
JOHNSON, P. and PERRELLA, M. (1971). *Proc. R. Soc. B.* **176**, 445. [131]
JULLANDER, I. and SVEDBURG, T. (1948). *Nature*, **153**, 523. [85]
KATCHALSKY, A., ALEXANDROWICZ, Z., and KEDEM, O. (1966). In *Chemical physics in ionic solutions* (ed. B. E. Conway and R. G. Barradas), p. 295, Wiley, New York. [54]
—— and SPANGLER, R. (1968). *Quart. Rev. of Biophys.* **1**, 127. [63]
KAWAHARA, K. and TANFORD, C. (1966). *Biochemistry* **5**, 1578. [113]
KUHN, W. H. (1951). *Elektrochem. Angew. Physik. Chemi.* **55**, 207. [85]
KUPKE, K. and LINDERSTRØM-LANG, K. (1954). *Biochem. Biophys. Acta* **13**, 153. [122]
—— (1960). *Adv. Prot. Chem.* **15**, 57. [2, 55]
LAPANJE, S. and TANFORD, C. (1967). *J. Am. Chem. Soc.* **89**, 5030. [116, 117]
LAUFFER, M. A. (1966). *Biochemistry* **5**, 1952. [83]
LAURENT, T. and OGSTON, A. G. (1963). *Biochem. J.* **89**, 249. [62, 118]
LEE, T. D. and BERNS, D. S. (1968). *Biochem. J.* **110**, 457, 465. [126]
LOEB, M. (1922). *Proteins and the theory of colloidal behaviour.* McGraw-Hill, New York. [2]
LUCK, J. M., COOK, H., ELDREDGE, N. T., HALEY, M. I., KUPKE, D. W., and RASMUSSEN, P. S. (1956). *Arch. Biochem.* **65**, 449. [131]
MCMILLAN, W. G. and MEYER, J. E. (1945). *J. chem. Phys.* **13**, 276. [11]
MANNING, G. S. (1969). *J. chem. Phys.* **51**, 924. [54]
MASSON, C. R. and MELVILLE, H. W. (1949). *J. Polymer. Sci.* **4**, 323. [74, 91]
MEYER, F. A., COMPER, W. D., and PRESTON, B. N. (1971). *Biopolymers* **10**, 1351. [54]
MORAWETZ, H. (1965). *Macromolecules in solution*, Interscience, New York. [2, 52]
NICHOL, L. W., BETHUNE, J. L., KEGELES, G., and HESS, E. L. (1964). In *The proteins* (ed. H. Neurath), vol. II ch. 9. Academic Press, New York and London. [59]
NICHOL, L. W., OGSTON A. G., and PRESTON B. N. (1967). *Biochem. J.* **102**, 407. [39]
OGSTON, A. G. (1937). *Biochem. J.* **31**, 1952. [118]
—— (1962). *Arch. Biochem. Biophys. Suppl.* **1**, 39. [119]
—— (1966). *Fed. Proc.* **25**, 1112. [88]
—— and PRESTON, B. N. (1966). *J. Biol. Chem.* **241**, 17. [119]
—— —— (1973). *Biochem. J.* **131**, 843. [88, 90]
—— and WELLS, J. D. (1970). *Biochem. J.* **119**, 67. [88]
OLMSTEAD, E. G. (1966). *Mammalian cell water.* Henry Kimpton, London. [64, 103]
OVERBEEK, J. T. G. (1956). *Progress in Biophysics* **6**, 57. [17, 22]
PAGLINI, S. (1968). *Anal. Biochem.* **23**, 248. [83]
—— and LAUFFER, M. A. (1968). *Biochemistry* **7**, 1827. [123]

REFERENCES

PATTAT, F. (1959). *Makromol. Chemi.* **34**, 120. [90]
PEACOCKE, A. R. and PRESTON, B. N. (1960). *Proc. R. Soc. B.* **153**, 90. [56]
PHILIPP, H. H. and DJORK, C. F. (1951). *J. Polymer. Sci.* **6**, 383. [90]
PFEFFER, E. and DE VRIES, J. (1889). *Z. physik. Chem.* **3**, 103. [1]
PORTZEHL, H. (1950). *Z. Naturforsch.* **56**, 75. [52]
PRESTON, B. N., DAVIES, M., and OGSTON, A. G. (1965). *Biochem. J.* **91**, 449. [62, 119]
—— and MEYER, F. A. (1971). *Biopolymers* **10**, 35. [54]
PRESTON, B. N., and SNOWDEN, J. M. (1972). *Biopolymers*, **11**, 1627. [54]
——, ——, and HOUGHTON, K. T. (1972). *Biopolymers* **11**, 1645. [54]
PROCTOR, H. R. and WILSON, J. A. (1916). *J. chem. Soc.* **109**, 307. [88]
REID, E. W. (1904). *J. Physiol.* **31**, 438. [2]
—— (1905). *J. Physiol.* **33**, 12. [2]
REES, E. D. and SINGER, S. J. (1955). *Nature* **176**, 1072. [122]
RENKIN, E. M. (1954). *J. Gen. Physiol.* **39**, 225. [90]
ROLFSON, F. B. and COLL, H. (1964). *Analyt. Chem.* **36**, 888. [82]
ROWE, D. S. and ABRAHAMS, M. E. (1957). *Biochem. J.* **67**, 431. [75, 111]
SCATCHARD, G., BATCHELDOR, A. C., and BROWN, A. (1946). *J. am. chem. Soc.* **68**, 2320, 2315. [3, 14, 17, 25, 47, 74]
——, GEE, A., and WEEKS, J. (1954). *J. phys. Chem.* **58**, 783. [3]
—— and PIGLIACAMPI, J. (1962). *J. am. chem. Soc.* **14**, 127. [120]
——, SCHEINBERG, I. H., and ARMSTRONG, S. H. (1950). *J. am. chem. Soc.* **72**, 535, 540. [48]
SCHACHMAN, H. K. and EDELSTEIN, S. J. (1966). *Biochemistry* **5**, 2681. [113]
SCHULTZ, G. V. and KUHN, W. H. (1961). *Macromol. Chemie.* **50**, 37. [85]
STAUFFER, H., SRINIVASAN, S., and LAUFFER, M. A. (1970). *Biochemistry*, **9**, 193. [123]
STEINER, R. F. (1954). *Arch. Biochem. Biophys.* **49**, 400. [59]
—— (1970). *Biochemistry* **9**, 1375. [62]
—— (1968). *Biochemistry* **7**, 2201. [62]
SMITHIES, O. (1953). *Biochem. J.* **55**, 57. [75]
SMOLUCHOWSKI, M. von. (1918). *Z. Phys-Chem.* **92**, 129. [123]
SØRENSEN (1917). *C. r. Lab. Carlsberg.*, **12**, 1. [2]
STARLING, E. H. (1899). *J. Physiol.* **24**, 317. [2]
STAVERMAN, A. J. (1951). *Recueil.*, **70**, 344. [63]
TANFORD, C. (1961). *Physical chemistry of macromolecules*, Wiley, New York. [2, 17, 22, 27, 28]
—— (1968). In *Solution properties of natural polymers*, Chem. Soc. Special Publication, No. 23, p. 1. [55, 114]
——, KAWAHARA, K., and LAPANJE, S. (1966). *J. Biol. Chem.* **241**, 1926. [55]
—— (1967). *J. am. chem. Soc.* **89**, 729. [52, 55, 114]
TELLER, D. C., KINKADE, J. M., and COLE, D. R. (1965). *Biochem. Biophys. Res. Comm.* **20**, 739. [131]
THAIN, J. F. (1967). *Principles of osmotic phenomena*. Royal Institute of Chemistry Monographs for Teachers. [2]
TOMBS, M. P. (1970). In *Proteins as human food* (ed. R. A. Lawrie). Butterworths, London. [123]

TOMBS, M. P. and LOWE, M. (1967). *Biochem. J.* **105,** 181. [55, 76, 113]
TOMPA, H. (1956). *Polymer solutions.* Butterworths, London. [2]
VAN'T HOFF, H. H. (1888). *Phil. Mag.* **26,** 81. [1]
VAUGHAN, M. F. (1958a). *J. Polymer Science.* **33,** 417. [90]
—— (1958b). *Nature* **182,** 1730. [90]
—— (1959). *Nature* **183,** 43. [90]
VISSER, J., DEONIER, R. C., ADAMS, E. T., and WILLIAMS, J. W. (1972). *Biochemistry* **11,** 2634. [62]
WEISSBERGER, A. (ed) (1959). *Physical methods of organic chemistry.* Interscience, New York. [2, 45]
—— (1961). *Physical methods of organic chemistry.* Interscience, New York. [36]
WELLS, H. S. (1932). *Am. J. Physiol.* **101,** 409. [91]
WILLIAMS, J. W., VAN HOLDE, K. E., and BALDWIN, R. L. (1958). *Chem. Review.* **58,** 715. [16, 17]
WOLF, A. W. (1966). *Aqueous solutions and body fluids.* Harper & Row, New York. [102]
ZIMM, B. H. (1946). *J. chem. Phys.* **14,** 164. [28, 51]

INDEX

Acetobacter Xylinium
    production of membranes by, 91
activity coefficients
    definition of, 10
    relation to osmotic coefficients, 13
    relation to virial coefficients, 12
acrylamide gels
    use of as osmometers, 90
Adair
    on haemoglobin, 104
    simple osmometer, 69
aggregation, estimation of
    of albumin, 123
    of glycinin sub-units in urea, 101
associating systems, *see* dissociating systems
albumin (serum)
    mixtures with globulins and Johnston–Ogston effect, 118
    mixtures with haemoglobin, 129
    mixtures with hyaluronic acid, 118
aldolase
    sub-units, 113
alcohol dehydrogenase
    molecular weight, 115
arachin
    sub-units, 114
    dissociation as function of urea concentration, 114
axial ratio
    calculation from virial coefficient, 51

B, *see* second virial coefficient
binding coefficients
    definition of, 16, 18, 19
blood proteins
    early studies of osmotic pressure of, 2
boiling point elevation, 31, 36
bovine serum albumin
    aggregation of, 123
    molecular weight of, 111, 115
    specific volume of, 116
    virial coefficient of, 111, 115

C, *see* third virial coefficient
calculation
    of axial ratios from virial coefficients, 51
    detailed example of, (glycinin), 91
    of molecular weights, 92, 109, 111
    of sub-unit numbers, 93, 113
    of virial coefficients, 93, 111
Casassa and Eisenberg definition of components, 18ff
cell membranes, leaks in, 63
cellulose acetate membranes, 91
charge–charge interactions
    albumin–haemoglobin, 120
    soybean proteinase inhibitor, 127
charged macromolecules
    in ideal, osmotic (Donnan) equilibrium, 17
    in multi-component systems, 45
    in two-component systems, 40
    in three-component systems, 41
    mixtures of, 45
classification of osmometers, 66ff
colligative methods, 7, 31
    number average molecular weights from, 45
colloids
    osmotic pressure of, 1
collodion, (cellulose nitrate)
    use of for membranes, 91
concentration
    effect of errors in, 95, 111
    scales, 4, 26, 45
components
    definitions of, 18, 20, 44

Definition
    of macromolecular components by Casassa and Eisenberg, 18ff
    by Scatchard, 20ff
    of osmotic coefficient; 10, 13, 34
    of pressure units, 101
    of thermodynamic quantities, 3ff
denatured proteins
    intrinsic viscosity of, 115
    sedimentation coefficient of, 115
    specific volumes of, 116
    sub-unit number of, 114 ff
    virial coefficients of, 115
deuterium oxide
    effect on dissociating systems, 126
deviations from van't Hoff equation, 35
dextran beads
    use as osmometers, 89

INDEX

dissociating solvents
    acetic acid, 124
    dimethyl formamide, 122
    guanidinium hydrochloride, 114
    formamide, 112
    urea solutions, 114
dissociating systems
    examples of, 120ff
    histones, 132
    ideal, 58
    insulin, 120
    non-ideal, 60
    osmotic pressure of, 55ff
    phycocyanin, 127
    tobacco mosaic virus protein, 123
Donnan equilibria
    in ideal solutions, 17, 23
    in non-ideal solutions, 123
    magnitude of in haemoglobin, 105
dynamic osmometers, 66, 73

Electronic osmometers, 75
enolase, molecular weight, 115
entropy
    of mixing, 8
    non ideality and, 49
    second virial coefficient and, 50
    of TMV protein association, 124
errors
    from osmometer construction, 79ff
    general statistical analysis for, 97ff, 108ff
    filtration, 83
extrapolation
    to zero concentration, 30
    methods of, 111
    difficulties of, 121
    to determine equilibrium pressures, 73

Filtration errors, 83
free energy of association of
    histones, 134
    soya bean proteinase inhibitor, 127
    TMV protein, 124
freezing point depression, 31, 36
Fuoss–Mead osmometer, 71

Gibbs–Duhem equation, 6
gel
    shrinkage, osmotic, 88
gel osmometers, 88, 90
glycinin
    osmotic pressure of, 92ff
guanidinium hydrochloride solution
    effects on proteins, 114ff

haemocyanin
    molecular weight of, 106

haemoglobin
    derivatives of, 128
    hybridization of, 128
    magnitude of Donnan effect, 105
    mixture with albumin, 129
    molecular weight of, 2, 104
histones
    as associating systems, 133
    molecular weights of, 133
    non-ideality of, 132
    virial coefficients of, 134
hyaluronic acid, 188
    mixtures with albumin, 118
hybridization of haemoglobins, 128

ideal solutions
    chemical potential of, 6ff
    mixing of, 8ff
    Donnan equilibria in, 17, 23
    osmotic pressures of, 7
imperfect membranes, 62
insulin
    molecular weight of, 120
    dissociation of, 120
interchain link frequencies, 101
isotonic, definition of, 103

keratin
    osmotic pressure of, 112

lactate dehydrogenase
    molecular weight of, 115
    specific volume of, 116
light scattering, relation to osmotic pressure and sedimentation equilibrium, 27
low-pressure osmometers, 85
lysozyme, molecular weight of, 112

membranes, 1, 90ff
membrane potentials, 43, 104
molecular weights
    number average, 45, 93
    collected values of, 114
    calculation from experimental results, 93ff, 111
mixtures of non-reacting proteins, 117ff
multi-component systems, theory of, 45ff

neutral macromolecules
    in two component systems, 15, 34
number average molecular weights
    definitions of, 45
    use of sub-unit numbers in calculation, 99, 113
    comparison with weight average and Z average, 45, 113

INDEX

Osmolality, osmolarity
   definitions of, 102
Osmotic
   coefficient, 10, 13, 34
   balance, 86
   equilibrium in terms of chemical potentials, 32
osmotic pressure
   definition of, 2, 29
   historical significance of, 2
   meaning of, 2, 30ff
   origins of, in solvent transport, 28, 36
   rigorous equations for, 37ff
osmometers
   classification of 66ff
   design of 66ff
   defects in, 79ff
   dynamic, 31, 66, 73
   electronic, 75ff
   gel particle, 88ff
   low pressure, 85ff
   simple, 67, 69
   static, 31, 66
ovalbumin
   comparison with plakalbumin, 108ff
   molecular weight of, 107, 115
   specific volumes of, 116
   virial coefficients of, 115

Partial molar quantities, 5
Pascal
   definition of, 101
   magnitude of, 101
phycocyanin
   molecular weight of, 127
   association of, 128
plakalbumin
   molecular weight of, 107
precision of osmometry, 111
pressure units, 101
protein concentration, effect of errors in, 100, 107, 111

quaternary systems (salt–water–hyaluronic acid–albumin mixtures), 119

R, gas constant, 36
   values for, 97
random coils, 117
reflection coefficient, 63

Scatchard definition of components, 14ff
second virial coefficient
   associating systems, 55
   calculation of, 97ff
   factors determining
     charge, 18, 47
     exclusion effects, 49ff
     hydration, 53, 125
     size and shape, 48
   forms of, 12
   significance of, 12, 18, 46ff
   units of, 98
   values of, for native and denatured proteins, 115
serum albumin, see albumin and bovine serum albumin
shrinkage, and swelling of gels, 89
   of mammalian cells, 64
solvent,
   transport, rate of, 68, 73, 78, 84
soya bean proteinase inhibitor
   association of, 126
   charge–charge interactions in, 127
specific volume
   values of, in dissociating solvents, 114
   from osmometry and sedimentation equilibrium, 116
   contribution to virial coefficient, 51, 52, 100
statistical analysis of results, 93ff, 111
sucrose, osmotic pressure of, 2

Temperature effects
   on osmotic pressure, 7, 69
   in dissociating systems, 124
ternary systems, 119
third virial coefficient, 12
   factors affecting, 54
tobacco mosaic virus protein, 123ff

Units
   of pressure, 101
   of virial coefficients, 98
   of RT, 97, 98, 101

van't Hoff's osmotic pressure equation, 2, 34, 40, 102
virial coefficients,
   alternative forms of, 12
virial expansion, of osmotic pressure equation, 11, 27
   relation to equation of state for gases, 50
viscosity
   filtration errors, 83
   intrinsic, and protein configuration, 114